Case Studies in Information Warfare and Security

For Researchers, Teachers and Students

Edited by

Matthew Warren

Case Studies in Information Warfare and Security
Volume One
First published: April 2013

ISBN: 978-1-909507-06-7

Disclaimer: While every effort has been made by the editor, authors and
the publishers to ensure that all the material in this book is accurate and
correct at the time of going to press, any error made by readers as a result
of any of the material, formulae or other information in this book is the
sole responsibility of the reader. Readers should be aware that the URLs
quoted in the book may change or be damaged by malware between the
time of publishing and accessing by readers.

Note to readers.
Some papers have been written by authors who use the American form of
spelling and some use the British. These two different approaches have
been left unchanged.

Published by: Academic Conferences and Publishing International Limited,
Reading, RG4 9SJ, United Kingdom, info@academic-publishing.org
Printed by Lightning Source POD

Available from www.academic-bookshop.com

Contents

List of Contributors

Kari Alenius, *Department of History, University of Oulu, Finland*

Christian Czosseck, *Cooperative Cyber Defence Centre of Excellence, Tallinn, Estonia*

Zama Dlamini, *Command and Control and Information Warfare, DPSS and CSIR, Pretoria, South Africa*

Steven Furnell,' *University of Plymouth UK and Edith Cowan University Perth Australia*

Marthie Grobler, *Council for Scientific and Industrial Research, Pretoria, South Africa*

Shona Leitch, *Deakin University, Australia*

Mapule Modise, *Command and Control and Information Warfare, DPSS and CSIR, Pretoria, South Africa*

Rain Ottis, *Cooperative Cyber Defence Centre of Excellence, Tallinn, Estonia*

Julie Ryan, *The George Washington University, Washington, USA*

Daniel Ryan, *National Defense University, Washington, USA*

Anna-Maria Talihärm, *Cooperative Cyber Defence Centre of Excellence, Tallinn, Estonia*

Joey Jansen van Vuuren, *Council for Scientific and Industrial Research, Pretoria, South Africa*

Allen Wareham, *University of Plymouth UK*

Matthew Warren, *Deakin University, Australia*

Jannie Zaaiman, *University of Venda, South Africa*

Introduction to Case Studies in Information Warfare and Security

Modern society is highly dependent on key critical systems either physical or technology based. They have become more significant as the information age has developed and societies have found themselves dependant on these systems. These key systems are grouped together and are described as critical infrastructures; which can be at risk of Information Warfare or cyber attacks. The threat of Information Warfare or cyber attacks to governments and businesses critical infrastructures has become a more pressing and everyday issue facing countries all around the world. Information Warfare causes many separate and individualised problems and the perpetrators can be a country, sub-state groups or individuals which makes defence of such attacks even more troublesome. One particularly difficult aspect of protecting against Information Warfare is the sheer number of possible attack types that could occur against critical infrastructure systems and the fact that passive attacks may occur rather than active attacks, for example, monitoring network trend data rather than trying to disrupt a network.

It is hard for us as individuals to imagine the consequence of an Information Warfare attack: an extended loss of power or the failure of related systems such as banking systems, the Internet, lifesaving medical equipment, the failure of public transportation systems, water treatment plants being non-functional or a lack of food at the supermarkets due to the malfunction of food distribution systems. It is because of these potential Information Warfare

risks that governments in all countries are so concerned and taking action against these new security threats.

Many critics dismiss the cyber threat to countries as being "hype
or overstated", but that is far from the truth. In 2007, cyber attacks on Estonia resulted in the failure of Estonia's online infrastructure; in May 2007 we saw that Estonia was the first victim of
a prolonged Information Warfare campaign. The background to
the campaign was a political disagreement between Estonia and
Russia that was taking place in the physical world. In 2010, we
saw the development of the Stuxnet malware (in this case a
worm) that had the ability to disrupt certain types of SCADA (Supervisory Control And Data Acquisition) systems that support key
industrial systems, such as power supplies and water treatment
facilities. The development of the Stuxnet malware raised many
security concerns about SCADA systems. In many cases these
SCADA security concerns could be mitigated by something as simple as implementing an effective patch management system. The
Stuxnet example however highlights the recent capability and
complexity of malware and the possibility for malware to be used
as a weapon in an Information Warfare campaign. The increasing
complexity of these malware attacks could make it harder for
governments and organisations to protect their critical infrastructures against these particular threats. This is amplified by the zero
day attack, where the malware appears for the first time online
and there is no suitable protection against the attack; the time
needed to analyse the malware and develop the suitable protection signature means that any attempts to mitigate are futile.

In Australia 438 cyber incidents occurred between 2011-12 which
required a significant response by the Australian Government Cyber Security Operations Centre; in 2012, the hacking group
Anonymous stole personal information of hundreds of thousands
of customer from an Australian ISP and then disclosed that information online. Countries have developed unique approaches to
deal with the risks of cyber attacks. In Australia, the Prime Minis-

ter Julia Gillard released Australia's first *National Security Strategy, Strong and Secure: A Strategy for Australia's National Security.* This strategy reinforced the importance of the protection of Australian's against many security threats including cyber threats. The Australian government's strategy acknowledged the threat of cyber espionage and foreign interference and the threat to "classified government information; commercial information with direct consequences for business and the economy; intellectual property; and the private information of Australian citizens". This is a significant acknowledgement that cyber threats now impact every Australian and have become an issue not only for governments but for individuals and their online information. The need to "strengthen the resilience of Australia's people, assets, infrastructure and institutions" against cyber attacks has also been recognised. This means the issue is not just one of protecting against cyber attacks, but also the ability to rebuild systems quickly after a cyber attack and minimise their impact. Tied up with the new strategy was the announcement of the creation of a new Australian Cyber Security Centre which will be in operation by the end of 2013 and which aims to improve partnerships between government and industry. The centre will have the ability to protect against new and developing cyber security threats in real time and allow for information to be shared quickly, so any cyber risks can quickly be mitigated.

The Australian Government has also identified that resilience is an important issue for critical systems. The Australian Government focuses on resilience from the following aspects in terms of (Critical Infrastructure Resilience Strategy, 2010):

Critical infrastructure resilience - the ability to reduce the magnitude, impact or duration of a disruption to critical infrastructure whatever its cause. So if essential services are damaged or destroyed, they can get up and running again quickly. This is an important part of creating a nation where all Australians are better able to adapt to change, have reduced exposure to risks, and are better able to bounce back from disaster;

Mutual responsibility – the responsibility of critical infrastructure resilience management and operation is shared between the owners and operators of critical infrastructure, and all levels of government – Australian, state and territory, and local. The owners and operators of critical infrastructure are primarily responsible for ensuring the security of their assets.

Smaller countries such as New Zealand also have formed dedicated critical infrastructure protection centres to secure against cyber risks but it was the USA that was the first to recognise the significance of critical infrastructure and the importance of defending these critical infrastructures. They were the first to implement comprehensive measures to protect US critical infrastructures. Since those first measures in the late 1990s, the US government has implemented further measures recently announcing the expansion of US Defense Department's Cyber Command from 900 to more than 4,000 staff. The US government is planning three different forces under Cyber Command: "national mission forces" to protect computer systems that support the nation's power grid and critical infrastructure; "combat mission forces" to plan and execute attacks on adversaries; and "cyber protection forces" to secure the Pentagon's computer systems. This represents a major step forward in cyber protection and a model that many other countries will likely reproduce over time. Many governments however have to consider whether the protection of critical infrastructure is considered a military, government, law enforcement or corporate issue or whether it be the work of an agency or department that coordinates and integrates some or all of these bodies.

One of the challenges of Information Warfare is the global nature of the threat. What we have seen is the need for international cooperation to deal with cyber threats, particularly in the area of sharing information, of countries being part of joint training exercises, and jointly coordinating cyber defence responses. We have witnessed Cyber security co-operation defined as part of the Australian – US defence treaty; and in response to the Estonia Information War we have seen the NATO Cooperative Cyber Defence Centre of Excellence be developed in order to protect NATO members against the threats of Information Warfare as well as joint statements on cyber security co-operation from both the United Kingdom and New Zealand governments. Many other governments have come to a similar conclusion; that they have a duty of care to protect their population and citizens against Information Warfare threats, but in times of austerity the consideration is whether governments are able to fund the required budgets for these key initiatives. The NATO Cooperative Cyber Defence Centre of Excellence could be an example of how budgets and resources could be shared between a number of countries to defend against Information Warfare risks. Whilst funding is important, the human resource consideration is essential and national strategies and approaches are needed. The UK government has recognised the shortage of Cyber Security specialists and introduced a number of initiatives to counteract this such as education programs about Cyber Security in school curriculums, the setting up of the *Cyber Security Challenge UK* to providing advice, support and guidance for anyone interested in a career in cyber security and the development of the *UK Certification for Information Assurance Professionals* to accredit Cyber security professionals.

The aim of this book is to highlight a number of important and significant cases in relation to Information Warfare. The papers come from authors from the following continents: Africa; Europe; Oceania and North America and includes papers from the European Conference of Information Warfare and the International Conference of Information Warfare. A number of chapters focus on the Estonian Information Warfare Attacks, considering the reasons behind their occurrence, outlining what occurred, those who were involved and the important lessons to be drawn from their experiences. The book assesses how individual countries deal with Information Warfare in terms of protecting critical infrastructures or raising security awareness amongst a population and reflects on other considerations of Information Warfare in terms of the neutrality in Cyber Warfare, co-operation and the role of activism.

Professor Matthew Warren
School of Information Systems
Deakin University
Australia
www.mjwarren.com

About the Author

Matthew Warren is a Professor of Information Systems at Deakin University. He is the former Head of School of the School of Information Systems (for a period of eight years) at Deakin University.

Professor Warren is a researcher in the areas of Information Security, Computer Ethics and Cyber Security.

He has authored and co-authored over 300 books, book chapters, journal papers and conference papers. He has received numerous grants and awards from national and international funding bodies, such as: Australian Research Council (ARC); Engineering Physical Sciences Research Council (EPSRC) in the UK; National Research Foundation in South Africa and the European Union. Professor Warren regularly reviews research proposals submitted to the Australian Research Council and the South African National Research Foundation.

Professor Warren gained his PhD in Information Security Risk Analysis from the University of Plymouth, United Kingdom and he has taught in Australia, Finland, Hong Kong and the United Kingdom.

Cyber Security Awareness Initiatives in South Africa: A Synergy Approach

Zama Dlamini and Mapule Modise
Command and Control and Information Warfare, DPSS and CSIR, Pretoria, South Africa
Originally Published in the Conference Proceedings of ICIW 2012

Editorial Commentary

The paper focuses on the Technological advances have changed the manner in which ordinary South African citizens conduct their daily activities. Many of these activities are carried out over the Internet. As a response, various entities engage in cyber security awareness initiatives and trainings with the aim to create cyber security awareness (CSA) among the citizens of South Africa. In the absence of a national cyber security policy, however, these awareness initiatives and programmes are delivered through a variety of independent mechanisms. Various entities engage in cyber security awareness training each with its specific objectives and focus areas. It is argued in this paper that cyber security is complex and multi-faceted. No single solution can effectively address it. While the current means to create cyber security awareness does make impact, the fragmented and uncoordinated nature thereof have a potential to create its own dynamics. This paper evaluates the extent to which the current cyber security awareness initiatives address the cyber security threats and risks. The assessment is based on the initiatives objectives, alignment of the programme to the cyber threats, and the target audience.

Abstract: Technological advances have changed the manner in which ordinary citizens conduct their daily activities. Many of these activities are carried out over the Internet. These include filling tax returns, online banking, job searching and general socialising. Increased bandwidth and proliferation of mobile phones with access to Internet in South Africa imply increased access to Internet by the South African population. Such massive increased in access to Internet increases vulnerabilities to cyber crime and attacks and threatens the national security. As a result, South Africa remains one of top three countries that are targeted by phishing attacks, the other two are the US and the UK (RSA, 2011). As a response, various entities engage in cyber security awareness initiatieves and trainings with the aim to create cyber security awareness (CSA) among the citizens of South Africa. In the absence of a national cyber security policy, however, these awareness initiatives and programmes are delivered through a variety of independent mechanisms. Various entities engage in cyber security awareness training each with its specific objectives and focus areas. It is argued in this paper that cyber security is complex and multifaceted. No single solution can effectively address it. While the current means to create cyber security awareness does make impact, the fragmented and uncoordinated nature thereof have a potential to create its own dynamics. The focus of organisations to deliver on their own objectives translates to some extent into the optimisation of the behaviour of individual entities as opposed to the optimisation of the national cyber security awareness as a whole. This paper evaluates the extent to which the current cyber security awareness initiatives address the cyber security threats and risks. The assessment is based on the initiatives objectives, alignment of the programme to the cyber threats, and the target audience.

Keywords: national security, cyber security awareness, cyber fraud, cybercrime, cyber threats

1. Introduction

Security and protection of individuals and organisation against the fast growing dangers of the cyber crime remain one of the major challenges facing cyber security experts, scholars and politicians. Cyber crime is on the rise in South Africa (SAPS, 2011). The increase is the result of the increased bandwidth and the proliferation of smart phones which has widened access to Internet to the majority of South Africans (RSA, 2011). Sixteen percent of the cyber crime

2

victims were affected through their phones compared to only 10 percent globally. Malware and computer viruses made up the biggest portion of cyber crime in South Africa. Scams and phishing fraud made up the rest. The total net cost of cyber crime in South Africa is estimated at R10.9 billion (Ferrier Int., 2011), making up one percent of the global net cost of R2.9 trillion. Cyber security awareness is the first line of defence against cyber attacks.

In South Africa, cyber security awareness initiatives are delivered through a variety of independent uncoordinated mechanisms. Various entities are engage on cyber security awareness training each with their specific objectives and focus areas. The cyber security is complex and multi-dimensional. An effective approach is that which accommodates and integrates all the dimensions. Therefore, the effectiveness of the current initiatives to the delivery of cyber security awareness initiatives that are relevant needs to be evaluated. The study outlines, evaluates and assesses the relevance of current cyber security initiatives in addressing cyber security challenges facing South Africa.

To achieve its aims, the paper is structured as follows: Section 2 defines key concepts of Information Security field and subsequently identifies the most prevalent cyber crime in South Africa. This is followed by Section 3 which identifies and describes current cyber security awareness initiatives in South Africa with specific reference to key questions that must be addressed by any cyber security awareness programme, the goals and objectives of individual initiatives, the group that is targeted, and the delivery method. The analysis of the relevance and effectiveness of the initiatives is based on how well the initiative responds to the challenges, alignment between the target group most attacked and those targeted for the initiatives. Section 4 concludes and presents future works.

2. Definition of key concepts

This section discus some of the significant concept in Information Security field and this study, these include: cyberspace, cyber crime and cyber security.

2.1. Cyberspace

Cyberspace refers to a physical and non-physical terrain created by and/or composed of some or all of the following: computers, computer systems, networks, and their computer programs, computer data, content data, traffic data, and users (IST-Africa, 2011).

Here are some of the threats associated with Internet or cyberspace that make cyber crime complex and difficult to eradicate (SAPS, 2011):

- *Cyber attacks are indirect:* Through cyberspace, nation-states can perpetrate espionage; industrial spies can steal trade secrets; criminals can steal money; and militaries can disrupt command-and-control communications.
- *Cyberspace or Internet is everywhere:* Today all business activities including production, manufacturing, transportation, telecommunications is heavily dependent on the Internet. There is a drive in South Africa to integrate all its government services' systems, such as home affairs, SARS (South African Revenue Service) and commercial banks.
- *The Internet has no boundaries*: the natural structure of Internet is complex and not easy to manage. This requires cooperation amongst all nations as one leak can be a threat to innocent users, who are not even residing where the leak was started or targeted (DiGregory, 2000).
- *Anonymity:* Everyone can be anyone they choose to become when they are online. Despite age or race, Internet, its applications and services, it is easy to lie about anything; this

has resulted in high rate of children abduction, and women abuse all over the world (Fick, 2009).

Since the users of cyberspace span across all the layers of the society, so does cyber attacks and therefore a comprehensive and integrated approach is required for the security of the citizens of any country.

2.2. Cyber crime

The draft of South African National Cybersecurity Policy defines cybercrime as illegal acts, the commission of which involves the use of information and communication technologies (South African Government Gazette, 2010).

2.2.1. Cyber crime in South Africa

Globally the US is the top hosting country for phishing attacks, that is, two out of every three phishing attacks that are identified. The countries that have consistently been among the top five hosts over the last six months include the US, UK, Canada, Germany and South Africa (RSA, 2011).

Business Against Crime from South Africa indicated that incidents of commercial crime involving computers had risen by 13% to 61 690 per 100 000 people between 2007 and 2008 (Ferrier Int., 2011). In 2011 alone, 84% of the South Africans who responded to Norton survey said they have been victims of cyber crime (France24, 2011). Online Fraud Report (October 2011) noted that South Africa remains among the 'top 5 attacked countries in the world' in terms of phishing attack volume in September (RSA, 2011). In February 2011, an estimated 18,079 phishing attacks were discovered to be aimed at South African networks, which accounts to an 11% increase from January, this was for the first time in nearly a year, that the total number of phishing attacks in a single month reached over 18,000 especially in South Africa. This makes phishing attacks the major cyber threat to South Africans (SAPS, 2011), (RSA, 2011).

The Crime Report 2010/2011 by South African Police Service (SAPS) also noted a steady increase is commercial crime in South Africa (SAPS, 2011). It is worth noting that computer or cyber crime is not explicitly singled out in this report but is seen as part of commercial crime. Commercial crime refers to any offence against statutory provisions which customs are responsible for enforcing, committed in order to either avoid payment of responsibility duties on movements of commercial goods; or any restrictions applicable to commercial goods; or receive any repayments, subsidies or other disbursements to which there is no proper entitlement; or illicit commercial advantage injurious to principle and practice of legitimate business competition (World Customs Organization, 2011).

This definition is open and encompasses many other illegal activities, which are mainly not cyber crime. If cyber crime is not specified and rated with various crimes that the country experience, the continuous increasing rate of cyber crime should be expected. This is shown in Table 1.

Table 1: Commercial crime in RSA from 2003/2004 to 2010/2011 (SAPS, 2011)

Province	2004/ 2005	2005/ 2006	2006/ 2007	2007/ 2008	2008/ 2009	2009/ 2010	2010/ 2011	2011/ 2012
Eastern Cape	4 398	4 498	5 726	5 363	6 767	7 795	8 345	8 658
Free State	2 561	2 425	2 311	2 677	3 250	3 498	4 730	7 524
Gauteng	23 337	24 368	26 869	26 986	30 757	34 095	34 756	31 153
KwaZulu-Natal	8 441	8 270	10 613	10 794	12 970	13 775	15 276	13 681
Limpopo	1 984	1 950	2 316	2 367	2 827	3 008	3 162	3 861
Mpuma-langa	2 474	2 630	2 860	3 778	4 082	4 683	4 609	5 581
North West	2 130	2 204	2 332	2 713	4 460	5 147	4 481	4 423
Northern Cape	955	730	844	949	995	1 144	1 141	1 134
Western Cape	7 651	7 139	7 819	9 659	11 366	11 697	11 888	12 035
RSA	53 931	54 214	61 690	65 286	77 474	84 842	88 388	88 050

The Table shows that there was a slight decrease in number of re-ported cases of commercial crime from 55 869 to 53 931 for the years 2003/2004 and 2004/2005. The years 2006 – 2011 have shown a consistent increase. Furthermore, Gauteng and the hub of economic activity has consistently remained the province with the highest reported cases since 2003.

Cyber crime incidents increase in monetary value every day in South Africa. Here are some of the examples: The Road Accident Fund has been stolen via the use of key loggers, accounting to the value of R15 million; the ABSA bank electronic fraud incidents cost up to R30 million, Landbank's electronic incidents accounts to R150 millions and the South African Revenue Services (SARS) electronic fraud inci-dents to the value of R100 millions in the year 2010 (SA government gazette, 2010), (The New Age, 2011).

The cyber crime that remain on top of the list of South African ma-jor cyber attacks and threats remain to be phishing attacks, identity theft and monetary fraudulent on all levels of national society. Other cyber crime and security threats that have been experienced in South Africa includes: Adware, Botnet, Cyber bullying, Cyber stalking, Data Theft, Hacking, Hoax Email, Key logging, Malware, So-cial Engineering, Spam Spyware and Trojan Virus (ISG-Africa, 2011)

Although reduction of cyber crime received a special mention dur-ing his 2009 State of the Nation address, President J.Z. Zuma stated that "Amongst other key initiative, we shall intensify our efforts against cyber crime and identity theft, and improve systems in our jails to reduce repeat offending", cyber crime rate in South Africa continue to increase everyday and there are no proper structures yet to deal with it (Zuma, 2009).

2.3. Cyber security

Cyber security is defined as the collection of tools, policies, security concepts, security safeguards, guidelines, risk management ap-proaches, actions, training, best practices, assurance and technologies that can be used to protect the cyber environment, organization and user assets.

Although South Africa currently does not have a cyber security policy, it recognises the need for a policy that will reduce the vulnerability of cyberspace, prevention of cyber threats and attacks and the ability to recover swiftly from any attack. To date, there is only a draft of cyber security policy that was released in February 2010 and a national cyber security policy framework is still drafted. The next section therefore will outline key objectives of cyber security policies from various countries.

2.3.1. South African cyber security policy

The SA Cyber security policy is made out of six key elements or strategic objectives to:

- Facilitate the establishment of relevant structures in support of cyber security;
- Ensure the reduction of cyber security threats and vulnerabilities;
- Foster cooperation and coordination between government and private sector;
- Promote and strengthen international cooperation on cyber security (SA government gazette, 2010).

2.3.2. US cyber security policy

A US cyber security policy review team suggest that any complete national cyber policy must consider, at a minimum, the following elements:

- Governance: Encompasses US Government (USG) structures for policy development and coordination of operational activities related to the cyber mission across the Executive Branch.

8

- Architecture: Deals with performance, cost, and security characteristics of existing information and communications systems and infrastructures as well as strategic planning for the optimal system characteristics.
- Norms of Behaviour: Addresses elements of law, regulation, and international treaties and undertakings, as well as consensus-based measures.
- Capacity Building: Encompasses the overall scale of resources, activities, and capabilities required to become a more cyber-competent nation (Cyberspace Policy Review, 2011).

2.3.3. Kenya cyber security policy

Kenyan cyber security policy is not ready yet. Currently, the following objectives are considered:

- Collaboration between stakeholders;
- Develop relevant Policies, Legal and Regulatory frameworks
- Establish national CERT thus providing a Trusted Point of Contact (TPOC);
- Build Capacity: technical, legal and policy;
- Awareness creation is key;
- Research and development;
- Harmonization of Cybersecurity management frameworks at the regional level (at the very least) (Ngudi, 2010).

2.3.4. Mauritius cyber security policy

Mauritius is one of the few countries who have most of cyber security infrastructures in place, including response team, incident response team, the awareness portal the cybercrime prevention committee and cyber security strategy, which have the following as its objectives:

- National Awareness Programs and Tools

- Good Governance of Cyber Security & Privacy
- Harnessing the Future to Secure the Present
- Personal Cyber Security
- A holistic approach integrates many elements, (Carnegie Mellon CyLab, 2008).

Although national cyber security policies differ from country to country, the key objectives and the approaches are gearing towards the achievement of the similar goal with cyber security awareness as a common feature.

3. Cyber security awareness

Cyber security awareness (CSA) is the security training that is used to inspire, stimulate, establish and rebuild cyber security skills and expected security practise from a specific audiences (Ministry of Defence-Estonia, 2008). It used to promote and encourage Internet users to practise safety precautions, and train them on online defence methods. Furthermore, it equips these users with cyber security skills on all the aspects of cyber security so that not only the national network infrastructures are kept resilience to cyber attacks and threats, but also the users are well informed (Dlamini *et al.*, 2011).

Any cyber security awareness initiative should have a plan, clearly defined goals and objectives, expected results, delivery methods, risks, and methods to evaluation the initiative.

Peltier believes that in order for the cyber security awareness program to be successful, there are five key factors that need to considered and ensured (Peltier, 2005). These include:

- A clear process to take the message to the users or targeted audience in order to reinforce cyber security as a significant concept.

- Identification of the individuals who are responsible for the implementation of cyber security awareness program.
- Determination and evaluation of the sensitivity of information and the criticality of cyber security infrastructure, applications and systems.
- The reasons for the implementation of cyber security concepts and awareness programs in convincing the audience of the significance of cyber security awareness programs that must be implemented.
- Ensuring that the related government department or the management supports the goals and objectives of the cyber security awareness program for the community.

Therefore, any cyber security awareness initiative has to answer a number of key issues if it is to be viable. Some of these issues may include:

- The programme /strategy which provide a coherent analysis of the state of cyber crime and security and the sources of these conditions. This analysis must of course identify the positions and interest of the different players within such a context.
- The initiative must identify the target group whose interest it seeks to address and specify the goals of such a group regarding cyber security awareness.
- The programme must specify the implementation plan and the how its effectiveness will to be measured.

3.1. Cyber security awareness initiatives in South Africa

The first part of this section identifies and describes the current Cyber Security Awareness (CSA) initiatives in SA. The research question for this part is *"do the initiatives/programmes contain the most basic requirement such as a plan, goals and objectives, delivery methods etc?"*

The second part evaluates the effectiveness of the initiatives/programme and the research question is *"Do the initiatives address the key issues making it viable?"*

3.1.1. Identification and description of CSA initiatives

A number of initiatives engaging in CSA have been identified. According to Grobler *et al.* (2011) the initiatives are spread across only four provinces (Gauteng, Mpumalanga, Limpopo, and Eastern Cape) out of the nine provinces in South Africa.

The list of these initiatives is summarized in Table 2 and Table 3. Table 2 demonstrates the initiatives' goal or objective/s, the targeted group and the topics included on training programmes.

Table 2: Cyber security awareness initiatives in South Africa

SA CSA Initiatives	Goal /Objective/s	Targeted Audience	Topics Discussed
CSIR (DPSS-CCIW)	To educate current and future users of the computers on safe and secure online habits To increase awareness and understanding of the dangers of the Internet. To provide individuals with the necessary knowledge to make the right decisions in Internet-related situations	Secondary schools, Further Education Training colleges Technical university students Non-technical university students Community centers Support staff Educators/Teachers	Physical Security, Malware and counter-measures, Surfing, Social aspects of cyber security
UP ICSA-PumaScope	Falls under an existing project is called PumaScope, and is the main focus of UP's cyber security awareness initiatives	Rural schools (children and adults), churches, orphan homes	Topic varies as it is community based project

12

SA CSA Initiatives	Goal /Objective/s	Targeted Audience	Topics Discussed
UNMM (ISM)	To educate the users about information security or, more specifically, to educate users about the individual roles they play in the effectiveness of one type of control, namely, operational controls.	General company end-users, entrepreneurs (public or private sectors), children, parents, tertiary and senior citizens.	Passwords management, information security principles and terminology, social engineering, phishing, desktop security, patch management, updating anti-virus software and backup procedures, email security,
UNISA	To contribute towards the creation of cyber awareness culture.	School children	Cyber security concepts
UFH	To improve performance in the areas in which an organisation has identified performance deficiencies. To test students' personal information security competency level because a user can be aware of an issue but not necessarily act on the knowledge that the person has gained.	University students (1st and 3rd year-levels)	Passwords management, information security principles and terminology, social engineering, phishing, desktop security, patch management, updating anti-virus software and backup procedures, email security,
SABRIC	To deliver measurable value to our clients through a team of energetic specialists who consistently provide high quality support services and products and, To contribute to the reduction of bank related crime through effective public private partnerships.	South African Banks' (SABRIC partners) Employees	Commercial fraud, online scams, device scams, online safety practise
ISG Africa	To drive awareness and education around information security risk and governance.	Organisations and society	Cyber threats and information security

SA CSA Initiatives	Goal /Objective/s	Targeted Audience	Topics Discussed
South African Centre for Information Security (CIS)	To develop and promote a management and governance drive to implement effective information security programs	Organisations in all aspects	Cyber crime

Table 3 illustrates the initiatives as well as their corresponding collaborations and their locations, the methods used by the initiatives to interact with target groups, the methods used to evaluate the programme and the outputs. Some of the organizations who provide such initiatives include:

Council for Scientific Industrial Research (CSIR) and University of Venda: The CSIR and the University of Venda are collaborating to raise cyber security awareness in local rural communities in the South African Limpopo province, Vhembe district. The motivation behind this initiative is to prevent innocent Internet users from becoming victims of cyber attacks, by educating novice Internet and technology users with regard to basic security (Grobler 2011).

University of Pretoria (UP) (ICSA- PumaScope): UP embarks on numerous community-based projects that serve many geographical areas in South Africa. An existing project is called PumaScope, and is the main focus of UP's cyber security initiatives awareness initiatives. It focuses on the transfer of knowledge in the areas of Computer Science, including basic computer literacy and cyber awareness (Grobler 2011).

Table 3: Cyber security awareness initiatives in South Africa

SA CSA Initiatives	Collaboration /Location	Delivery Methods	Evaluation Methods	Output
CSIR (DPSS-CCIW)	Univen, SAFIPA, Meraka, UP, Mpumalanga/ Pretoria	Presentation, Board and computer based games, Movie clips	Pre- and Post- Survey	Reports, Publication
UP ICSA-PumaScope	Mpumalanga, Gauteng, Microsoft, Atos origin/ Pretoria	Presentations	Surveys	Accredited certificates by the Department of Computer Science at the University of Pretoria.
UNMM (ISM)	Not mentioned/ Port Elizabeth	Computer Science portal, short learning Program, Module e-learning, competitions, gamming education	Survey	Degrees, paper publications, technical projects, games and content specific sites
UNISA	Not mentioned/ Pretoria	Schools' curriculum preparations and research	N/A	Curriculum, degrees and paper publication
UFH	Not mentioned/ East London	Formal lecture	Online questionnaires	lecture, textbook chapters and notes, Paper publication
SABRIC	Most South African banks/ Gauteng	Online discussion, web-portal and formal presentations	N/A	Reports
ISG Africa	Business against crime and SABRIC, South Africa and Africa/ South Africa	Web-portal and online discussions	N/A	N/A

SA CSA Initiatives	Collaboration /Location	Delivery Methods	Evaluation Methods	Output
South African Centre for Information Security (CIS)	Not mentioned/ South AfricaS	Discussions and presentations	N/A	Reports

University of Nelson Mandela Metropolitan (UNMM) (ISM): The Information Security Management research group at NMMU has been involved in research in the area of cyber security for more than a decade. The focus of the research group addresses all aspects of information security management, with information security awareness and related areas receiving a lot of attention. Many papers have been published and presented in these fields of information or cyber security awareness, culture and education (Grobler 2011).

University of Fort Hare (UFH): The Information Systems group at UFH conducted a study amongst a group of the university student within the campus who were non-IT. The focus of the group is to increase its capability in terms of keeping the computer users within the university aware of the threats and vulnerabilities that comes with the use of computers and Internet (Grobler, 2011).

University of South Africa (UNISA): The Information Security Awareness research group at UNISA works on a project that is mainly for children, adults and elderly. The initiative aims at teaching these groups on self responsibilities in securing their computers. (UNISA, 2011).

South African Banking Risk Information Centre (SABRIC): SABRIC was established as a wholly owned subsidiary of the Banking Association and is funded by most of South African banks. The centre has a public awareness initiative that is directly available on its website. The initiative focuses mainly on public needs and teaches public the best practises on the use of banks facilities and general security. It further gives tips on new scams and guidelines to follow if one becomes a victim (SABRIC, (2011).

Information Security Group of Africa (ISG-Africa): ISG-Africa consists of security professionals from corporate, government and IT / legal firms within Africa. The focus of this group is to establish, promote, manage and control various interest and user groups, for the promotion of education, and awareness of information security (ISG-Africa, 2011).

South Africa Centre for Information Security (CIS): CIS was established to develop and promote a management and governance drive to implement effective information security programs. CIS has a section where cyber crime is discussed and explained in details (SA-CIS, 2011).

Analysis of these initiatives shows that universities and research institutes such as the CSIR are at the forefront of the campaign to create cyber security awareness to the communities. There are some business organisations that have been identified but the details of how they conduct security awareness remain confidential. The authors understand that the list presented here is not exhaustive; these that are presented in this paper are those whose activities are on the public domain.

Most of the initiatives meet the basic requirements as all of them have plans, objectives, methods of delivery etc. Fifty percent of the identified initiatives have collaborative partnerships with other organisations and again fifty percent have methods to evaluate the initiative/programme. Furthermore, majority of the initiative make a huge contribution to the body of knowledge as seen by the number of publications, qualifications and reports.

3.1.2. Evaluation of effectiveness of cyber security awareness initiatives in South Africa

Table 4 below illustrates the evaluation of cyber security awareness initiatives against the key factors of cyber security awareness programme specified by Peltier (Peltier, 2005). The key factors includes the analysis of cyber crime, identification of target group, identification of the need of target group, evaluation of plans, evaluation methods and the involvement of the government.

Table 4: Evaluation of CSA initiatives

Initiatives	Does the initiative give a coherent analysis of the state of cyber crime?	Is the target group identified?	Are the needs of the identified group understood?	Are the plans implementable?	Are there Measures of Effectiveness?	Is the government involved?
CSIR (DPSS-CCIW)	Yes	Yes	Yes	Yes	Yes	No
UP (ICSA-PumaScope)	Yes	Yes	Yes	Yes	Not Specified	No
UNMM (ISM)	Yes	Yes	Yes	Yes	Yes	No
UNISA	Yes	Yes	Yes	Yes	Not Specified	No
UFH	Yes	Yes	Yes	Yes	Yes	No
SABRIC	Yes	Yes	Yes	Partially	Not Specified	No
ISG Africa	Yes	Not Specified	Not Mentioned	Partially	Not Specified	No
South African Centre for Information Security (CIS)	Yes	Not Specified	Not Mentioned	Partially	Not Specified	No

From Table 4, it is evident that all the initiatives identified address all the key issues of cyber security awareness with the exception of government involvement and measures of effectiveness. The limited or absence role of the government in these initiatives remains a huge concern and this together with the lack measures of effectiveness in most initiatives may hamper the roll out of cyber security awareness to wider communities.

4. Conclusion and future work

This paper presented some of the cyber security awareness initiatives in South Africa. These were analysed in terms of their target group, topics covered, collaboration, delivery methods, evaluation measures and the expected output. The initiatives were also evaluated against the key issues that any cyber security awareness initiative must tackle.

Analysis shows that the current initiatives are effective and have been able to address cyber security issues although at a smaller scale. This is despite the fact that South Africa is yet to develop a cyber security policy. Presently, universities are at the forefront with limited participation from the government. The business sector on the other hand seems to be engaging in cyber security awareness activities, though separately. A savvy cyber security aware nation requires that all levels of society be serviced. The initiatives are focusing largely on the communities. Literature material read for this paper indicates that these initiatives are comparable to the international counterparts.

It is therefore recommended that, a single body that will integrate all activities of all cyber security awareness initiatives is needed. It is envisaged that this body will develop a collaborated framework that spreads out all cyber security awareness across the country. Furthermore, this body can set standards procedures and measures for all its members to conform to. This body can also play the regulatory function to ensure that only quality cyber security awareness material is developed and presented to communities.

The formulation of the national cyber security awareness framework will be proposed in future, which will incorporate all the necessary roles and responsibilities of each initiative in order to ensure that the programmes reach all corners of the country.

References

Carnegie Mellon_ CyLab, (2008). "Message from Mauritius (Part II): Holistic Cyber Security Strategy -- Professional & Personal", available online from: http://www.gov.mu/portal/sites/csd/downloads/ppt/Track3/Mauritius.pdf, (accessed on 13/03/2011)

Cyberspace Policy Review, (2011). "Assuring a Trusted and Resilient Information", [online], http://www.whitehouse.gov/assets/documents/Cyberspace_Policy_Review_final.pdf, (accessed on 10/07/2011).

DiGregory, K., (2000). "Fighting Cybercrime-What are the Challenges Facing Europe?", Meeting before the European Parliament; United States Department of Justice [online], http://www.justice.gov/criminal/cybercrime/EUremarks.htm, (accessed on 08/01/11).

Dlamini, I.Z Taute, B. and Radebe, J., 2011. "Framework for an African Policy Towards Creating Cyber Security Awareness", [online], http://www.csir.co.za/dpss/docs/SACSAWFinal_16Aug.pdf, (accessed on 22 /09/2011).

Ferrier International, (2011). "Business Against Crime South Africa on the Latest Crime Statistics", [online], http://ferrierinternational.com/business-against-crime-south-africa-on-the-latest-crime-statistics/, (accessed on 05 /03/2011).

Fick, J., (2009). "Cybercrime in South Africa: Investigating and prosecuting cybercrime and the benefits of public-private partnerships", Council of Europe octopus interface conference cooperation against cybercrime, pp 10-1, (accessed on 08/07/11)

France 24, (2011). "Cybercrime costs $114 billion a year: Report", [online], www.physorg.com/news/2011-09-cybercrime-billion-year.html, accessed on (accessed on 16 /07/2011).

Grobler, M., Flowerday, S., von Solms, R. and Venter, H., (2011). "Cyber Awareness Initiatives in South Africa: A National Perspective", [online], http://www.csir.co.za/dpss/docs/SACSAWFinal_16Aug.pdf(accessed on 19/08/2011).

ISG-Africa, (2011). "CYBERCRIME: Safety & Security Guide", [online], http://cybercrime.org.za/local-resources/, (accessed on 25 /09/2011).

IST-Africa, (2011). "Proceedings of the First IFIP TC9 / TC11-Southern African Cyber Security Awareness Workshop 2011", [online], http://www.csir.co.zHYPERLINK "http://www.csir.co.za/dpss/docs/SACSAWFinal_16Aug.pdf"aHYPERLINK "http://www.csir.co.za/dpss/docs/SACSAWFinal_16Aug.pdf"/dpss/docs/SACSAWFinal_16Aug.pdf, (accessed on 23 /09/2011).

Ministry of Defence-Estonia, 2008. "Cyber Security Strategy- Cyber Security Strategy Committee", [online], http://www.kmin.ee/files/kmin/img/files/Kuberjulgeoleku_strateegia_2008-2013_ENG.pdf, (accessed on 27/08/2011).

Ngundi, V. (2010). Cybercrime, Cybersecurity and Privacy, available online from: http://www.eaigf.or.ke/files/2010_KIGF_Cybercrime_Cybersecurity_and_Privacy.pdf, (accessed on 05/03/2011)

Obama, B.H. 2009. Remarks By The President On Securing Our Nation's Cyber Infrastructure, BH Obama, President of the United States of America; The White House, Office of the Press Secretary, [online], http://www.whitehouse.gov/the_press_office/Remarks-by-the-President-on-Securing-Our-Nations-Cyber-Infrastructure/, (accessed on 13/03/2011).

Peltier, T., (2005) "Implementing an Information Security Awareness Program". Information Systems Security 14. Vol. 2, pp. 37–49.

RSA, (2011). "Cyber Security Awareness Month Fails to Deter Phishers", [online], http://www.rsa.com/solutions/consumer_authentication/intelreport/11541_Online_Fraud_report_1011.pdf, (accessed on 22/04/2011).

South African Government Gazette, (2010). "South African National Cyber Security Policy", [online], http://www.pmg.org.za/files/docs/100219cyber security.pdf, (accessed on 02/06/2011). ISSUE?

SABRIC, (2011). "Public Awareness", [online], https://www.sabric.co.za/?pg=Public+Awareness, (accessed on 02/06/2011).

South African Centre for Information Security, (2011). "What is cyber crime", [online], file:///I:/Attention/ICIW%202012/22%20November%202011/definecrime.htm, (accessed on 16 /09/2011).

South African Police Service, (2011). "Crime Report 2010/2011: South African Police Service (SAPS)", [online], http://www.saps.gov.za/statistics/reports/crimestats/2011/crime_situation_sa.pdf, (accessed on 19/08/2011).

21

The New Age, (2011). "SA a target for cyber crime", [online], http://thenewage.co.za/printstroy.aspx?news_id=28862&mid=53, (accessed on 22/09/2011).

UNISA, (2011). Information Security Awareness research group, [online], http://www.unisa.ac.za/default.html, (accessed on 12/08/2011).

Wolrd Cutoms Organization, (2011). "Terms of Reference for the Working Group on Commercial Fraud", [online], http://www.wcoomd.org/home_about_us_committstructcommrfraud.htm , (accessed on 16/07/2011).

Zuma, J.G. (2009). State of the Nation Address by His Excellency, JG Zuma, President of the Republic of South Africa; [online], http://www.parliament.gov.za/content/SONA3June2009.doc, (accessed on 27 /09/2011).

Neutrality in the Context of Cyberwar

Julie Ryan[1] and Daniel Ryan[2]
[1]The George Washington University, Washington, USA
[2]National Defense University, Washington, USA
Originally Published in the Conference Proceedings of ICIW 2011

Editorial Commentary

This paper will examine the legal antecedents of the concepts of neutrality and current enforceability of declarations of neutrality in the context of information operations amongst belligerents. This is a non-trivial point of understanding, given the potential for belligerents to use and abuse infrastructure elements owned and/or operated by nation states desiring to remain neutral.

The analysis will consider the instantiated concepts of neutrality, the potential for expanding or contracting the concepts of neutrality in the context of cyberwar, and the possibility of erosion of neutrality in cyberwar scenarios. The problem with cyberwar is that they are potentially not just transferring orders but also potentially weapons -- cyber-weapons. So it becomes a more complex problem and the challenge is to understand at what point the nation state should be required to act, or if such a point exists at all. This analysis of the paper will examine the intersection between technology and law in regards to this issue.

Abstract: This paper will examine the legal antecedents of the concepts of neutrality and current enforceability of declarations of neutrality in the context of infor-

23

mation operations amongst belligerents. This is a non-trivial point of understanding, given the potential for belligerents to use and abuse infrastructure elements owned and/or operated by nation states desiring to remain neutral. The analysis will consider the instantiated concepts of neutrality, the potential for expanding or contracting the concepts of neutrality in the context of cyberwar, and the possibility of erosion of neutrality in cyberwar scenarios. We have a notion enshrined in international law that says that you don't lose your neutrality if belligerents use your telephone lines or telegraph lines to communicate even if they are crossing your territory, even if they are passing operational orders. The problem with cyberwar is that they are potentially not just transferring orders but also potentially weapons -- cyber-weapons. So it becomes a more complex problem and the challenge is to understand at what point the nation state should be required to act, or if such a point exists at all. This analysis will examine the intersection between technology and law in regards to this issue.

Keywords: neutrality; law of armed conflict; international humanitarian law; cyberwar

1. War and the laws of armed conflict

During less than one percent of the last two million or so years of human evolution has agriculture and animal husbandry replaced the hunter-gatherer existence as a characteristic way of life. (Gat 2006, p. 4) During the hunter-gatherer phase, humans engaged in endemic primitive warfare. (Keegan 1193, p. 5 and pp. 115ff) As technology evolved, it influenced – and was influenced by – warfare, producing revolutions in military affairs. (Boot 2006, p. 8) The longbow, stirrups, gunpowder, conoidal bullets, machine guns, aircraft, radar, sonar, rockets and spacecraft, and now computers and precision-guided weapons, are but a small sample of the technologies that have continuously changed the face of warfare throughout history. As warfare became the province of nation-states, belligerencies between and among nations led to some states declaring their intent to remain neutral, and the development of conditions under which their neutrality was recognized by the belligerents and other conditions under which neutrality was lost. This paper addresses modern concepts of neutrality, and explores the potential for, and perhaps need to, change our concepts of neutrality in the context of

cyberwar as information technologies change warfare as it was previously practiced.

War is "a condition of armed hostility between States," (Hyde 1945,
p. 1686. Cited in Elsea & Grimmett 2007, p. 23) or "a contention,
through the use of armed force, between states, undertaken for the
purpose of overpowering another." (von Glahn 1992. p. 669. Cited
in Elsea & Grimmett 2007, p. 23) War is "an armed conflict, or a
state of belligerence, between two factions, states, nations, coali-
tions or combinations thereof. Hostilities between the opponents
may be initiated with or without a formal declaration by any of the
parties that a state of war exists." (Dupuy, p. 261) Marcus Tullius
Cicero (106-43 BCE) famously said in an oration, Pro Tito Annio
Milone ad iudicem oratio (Pro Milone), in defense of Titus Annius
Milo, who had been accused of murdering Publius Clodius Pulcher, a
political enemy, "Silent enim leges inter arma" (the law is silent in
times of war), (Clark 1907) but his assertion wasn't true in antiquity,
and isn't true today.

Except in limited conditions, war was made illegal by the Charter of
the United Nations, which is a treaty among the world's nations
signed in the aftermath of World War II, a terrible conflict in which
some fifty million (perhaps as many as eighty million) died worldwide. (White 2005) Article 2(4) of the Charter provides that, "All
Members shall refrain in their international relations from the
threat or use of force against the territorial integrity or political independence of any state, or in any other manner inconsistent with
the Purposes of the United Nations." However, Article 51 makes use
of military force is permissible in self-defense, and Article 42 makes
military force permissible if authorized by the Security Council.

When military force is used, its use is subject to other treaties that
limit the nature and extent of force that may be employed in
achieving military objectives. Philosophers, statesmen and military
commanders have struggled to balance the destructive forces of
armed combat with national and international humanitarian con-

cerns, (Kolb 1997, n. 3) leading to the twin concepts of jus ad bellum —"the conditions under which belligerents might justly resort to the use of armed force as a means of conflict resolution" (Hensel 2008, p. 5) — and jus in bello —"the conditions for the just employment of armed force at the strategic, operational and tactical levels during periods of armed hostilities" (Hensel 2008, p. 5) — that together comprise the notions of just war. The notion of jus in bello ("justice in war") was known to Sun Tzu in 4th century BCE China. (Giles) Even, so, the concept of jus in bello was more slow to develop than jus ad bellum. In addition to the United Nations Charter, limitations on the use of military force include inter alia the Geneva Conventions and Protocols, and the Hague Conventions.

2. Cyberwar

As human beings have moved into cyberspace, they have begun to engage in all the usual types of human behavior, good and bad, allowed by the technology: communicating, working, contracting, playing, and socializing, as well as stealing, breaching contracts, engaging in tortious behavior, and invading other users' privacy. Now, nation-states are looking at cyberspace as place to conduct warfare operations, and terrorists are examining the possibilities inherent in asymmetric attacks through cyberspace on critical infrastructures.

The "nature" of cyberspace, however, differs in significant ways from the physical, electrical, chemical, and photonic properties of "real" space. Communications across the Internet take the form of packets containing addressing and administrative data as well as the intended bits being exchanged. ("What is a packet?") The paths taken by packets exchanged across the Internet are under the control of algorithms within the switches that relay the packets. (Tyson 2001) The paths are neither known to nor controllable by the users of the network.

Traditional approaches developed in real space for responding to misbehavior are hampered in cyberspace by difficulties in attribution, and only a loose correlation exists between "location" in cyberspace and location of users and cyber equipment within traditional legal jurisdictions. These realities will certainly impact the development of weapons, strategies, doctrines and tactics for use in cyberwar and countering cyberterrorism. Nevertheless, nations will undoubtedly seek to exercise and enhance national power by means of information operations in cyberspace, and the laws of armed conflict that have served civilized nations well in real space must be examined to determine how they can be used, and if they must be changed, to meet the realities of cyberwar and cyberterrorism. This paper will specifically address the legal issues associated with nation-state neutrality as applicable to these new realities.

3. Neutrality during periods of belligerency

"Neutrality" refers to concepts in customary international law and treaty law concerning the non-participation of some nations in warfare when a state of belligerency exists among other nations. The laws of neutrality presuppose the coexistence of war and peace – belligerents and their allies at war with other belligerents and their allies, while diplomacy, commerce, communications and so forth continue with and among nations not involved in the belligerencies, both neutral states with other neutral nations and neutral states with the belligerents. (Neff 2000, p. 1. Cited by Kelsey 2008, p. 1442) Neutrality is a "legal, temporary situation of one state in relation to a conflict between two or more states. Neutrality consists in not participating directly in the war, through not rendering assistance to any belligerent party." (Osmanczyk & Mango 2004, A-F, p. 1547) It may be manifested by unilateral declaration or by entry into bilateral or multilateral treaties. Grotius identified two rules for neutrals: (1) neutrals should neither strengthen the position of a belligerent power with an unjust cause, nor hinder the position of a belligerent with a just cause, (Book III, Chapter XVII (III)(1))and (2) warring parties should be treated alike when the cause of the war is in doubt. (Book III, Chapter XVII (III)(1))

Even before the second half of the 19th century when the laws of war began to be codified in multilateral treaties, some principles relating to the conduct of armed hostilities had been included in bilateral treaties.... The rights and duties of neutrality in war, especially at sea, have been addressed in a large number of bilateral treaties between states from at least the early 17th century. [Footnote 12: W. E. Hall, The Rights and Duties of Neutrals, Longman's Green, London, 1874, pages 27-46, in a chapter surveying the growth of the law affecting belligerent and neutral states to the end of the 18th century, refers to "innumerable treaties" relating to neutrality that were concluded over several centuries (page 28).] Sometimes, following the conclusion of a bilateral treaty on neutrality, additional states proceeded [sic] to it. [Footnote 13: For example, on February 27, 1801 Denmark ceded to the convention between Russia and Sweden for the Reestablishment of an Armed Neutrality, which had been signed on 16 December 1800. 55 CTS (1799-1801) 411-24.] (Roberts & Guelff 1982, p. 4)

The law of neutrality was eventually codified in the Hague Conventions of 1907, including No. 3, Convention Relative to the Opening of Hostilities (requiring notice to neutrals of a state of war); No. 11, Convention Relative to Certain Restrictions with Regard to the Exercise of the Right of Capture in Naval War; and especially No. 5, Convention Respecting Rights and Duties of Neutral Powers and Persons in Case of War on Land. (The Avalon Project)

Having assumed a position of neutrality, a nation must not allow transit of military forces or equipment by belligerents across its land territory or the airspace above its land territory. The rules with respect to belligerent naval vessels, and aircraft flying over a neutral's territorial waters and exclusive economic zones, are more complicated. The notion of transit passage applies to "straits which are used for international navigation between one part of the high seas or an exclusive economic zone and another part of the high seas or an exclusive economic zone." (UNCLOS 1982, Art. 37) Ships and aircraft operated by belligerent nations may transit the territorial waters of a neutral state "solely for the purpose of continuous and expeditious transit of the strait..." (UNCLOS 1982, Art. 38) During transit passage, ships and aircraft must: "proceed without delay . . ., refrain from any threat or use of force against the sovereignty, territorial integrity or political independence of States bordering the strait . . ., and refrain from any activities other than those incident to their normal modes of continuous and expeditious transit unless rendered necessary by force majeure or by distress." (UNCLOS 1982, Art. 39)

The notion of innocent passage applies to passage through the territorial waters of a neutral state and is permitted "so long as it is not prejudicial to the peace, good order or security of the coastal State." (UNCLOS 1982, Art. 19) Passage is not innocent if it involves "any threat or use of force against the sovereignty, territorial integrity or political independence of the coastal State . . ., any exercise or practice with weapons of any kind, . . . any act of propaganda aimed at affecting the defence or security of the coastal State, . . . the launching, landing or taking on board of any aircraft [or] military device, [or] any act aimed at interfering with any systems of communication or any other facilities or installations of the coastal State."(UNCLOS 1982, Art. 19)

Once a state decides on a position of neutrality, it must take steps to prevent its territory from becoming a base for military operations of a belligerent. It must prevent the recruiting of military personnel, the organizing of military expeditions, and the constructing, outfitting, commissioning, and arming of warships for belligerent use. A neutral state is under no obligation to prevent private persons or companies from advancing credits or selling commodities to belligerents. Such sales are not illegal under the international law of neutrality. A neutral state may, if it chooses, go beyond the requirements of international law by placing an embargo upon some or all sales or credits to belligerents by its nationals. If it does so, it has the obligation to see that legislation, commonly referred to as neutrality laws, is applied impartially to all belligerents. Once enacted, neutrality laws are not to be modified in ways that would advantage one party in the war. (Neutrality 2008)

There is a limited communications exception in the law of neutrality for communications by belligerents and their allies across the land territory of neutral states. Hague Convention V, Article 8, provides, "A neutral Power is not called upon to forbid or restrict the use on behalf of the belligerents of telegraph or telephone cables or of wireless telegraphy apparatus belonging to it or to companies or private individuals." The Internet did not exist when the Hague Conventions were written, of course, but arguably this exception applies to Internet communications as well as telegraph and telephone communications. The nature and scope of this exemption is a key issue for neutrality in the context of cyberspace.

4. Neutrality in the context of cyberwar

When Hague V(8) was written, communications across the territory of a neutral nation via telegraph or telephone cables, or by wireless telegraphy, might have involved passing a variety of types of information. Command and control information might have been passed, for example, or intelligence or targeting information. Assuming that military units knew their own locations (not, necessarily, a reasonable assumption in those days), unit locations may have been reported. In short, information useful in prosecuting the belligerency, if it could be reduced to textual or numeric form suitable for transmission across the communications systems in use at that time, could be transmitted without imposing a burden on the neutral state to recognize or interdict the transmission. Some information may have been encoded or enciphered, and transmission would have necessarily been slow by today's standards, but fast relative to other media and transmission capabilities available at the time (foot, horseback, railroad, ship). (Lail 2002, p. 4)

Fast forward to the twenty-first century, and the ability to pass useful information across the Internet is much enhanced. Now not just text and numbers may be communicated, but sound to at least the level of voice recognition, imagery including high-quality color pictures, and measurement and telemetry data, such as GPS data, can be communicated quickly and easily across the Internet. Perhaps more importantly, tools and even weapons themselves, perhaps in the form of malware, can be moved across the territory of neutrals and belligerents alike using the Internet. Those engaged in such Internet communications do not and, for the most part cannot, know the path the packets comprising their communications will take, much less can they control the path. In fact, some of the packets may take different paths from other packets that are part of the same transmission, all transparent to and beyond the control of those engaged in the communication.

Historically, warfare has involved the use of kinetic weapons (e.g. projectiles) to kill and destroy. Modern warfare continues to use kinetic weapons, but may also use energy weapons – lasers, for example; but note that Protocol IV of the 1980 Convention on Certain Conventional Weapons specifically outlaws the use of blinding lasers – or may use logic weapons to attack and defend cyber-dependent infrastructures. In a modern warfare, information operations may be used in connection with kinetic operations (as in the confrontation between Russia and Georgia in 2008), (Tikk 2010, p. 66ff) or can be used without ancillary kinetic operations (as in the confrontation between Russia and Estonia in 2007). (Tikk 2010, p. 14ff) It is highly probable that we will never again see kinetic operations of any great extent without a cyber component. Whether information operations among nation-states without "armed conflict" will be deemed to be warfare probably depends upon the level of destruction realized. (Article 51 of the United Nations Charter uses the expression "armed attack" to justify war in self-defense by nation-states. However, the expression is not defined. It is not clear that it is proper, or desirable, to view a purely cyber incident as an armed attack. See Wingfield 2006, p. 12. See also Sullivan 2010) Information operations among, between or with non-nation-states cannot, by definition, be war, regardless of the level of destruction attained or the used of uniformed military personnel by one side or another and despite the common misuse of the term in referring to conflicts that are not between or among nation-states, as in "the global war on terror" (Rumsfeld Memo 16 October 2003) or the "war on drugs." (Testimony of OMB Director Nussle)

While belligerents' use of networks that cross a neutral's territory can take place without violating the neutrality status of the nations through whose territory the communications pass, Hague V(8) arguably did not foresee that that use might include weapons. The rules concerning neutrality require that passage of weapons or other military materials and equipment across the territory of a neutral must be interdicted by the neutral state, and if it fails to do so, or is unable to do so, the belligerents against whom the weapons or materials are to be used have a legal right to attack the transfer. (Brown 2006, p. 210) Hague V(1) forbids land transfers and Hague V(2) forbids use of the atmosphere. Some analysts have, therefore, concluded that cyberwar is not permitted under current neutrality law without a likely violation of the claimed neutrality. (Kelsey2008, pp. 1441-6) They recommend changes to bring the law into conformance with the reality of Internet transfers. (Kelsey 2008, pp. 1448-9) One recommendation would focus on intent: the rules of neutrality would not be violated unless the belligerent intended to use the information infrastructure of the neutral to deliver the weapons. The neutral would not have to interdict an unintentional passage, and would not be subject to attack by the other side based on an unintentional crossing of its territory by the cyber weapons. (Kelsey 2008, pp. 1448-9) This approach seems hopeless to us. The neutral probably has no knowledge that weapons are passing across its territory, could realistically do nothing if it did know, and has even less access to knowledge of the belligerent's intent with respect to the crossing.

However, there is an alternative approach to framing the problem and it's solution. Extra-atmospheric movements of weapons (other than nuclear weapons) and military materials above the territory of neutrals is permitted without imposing a duty on the neutral to interdict. The United Nations adopted a "Declaration of Legal Principles Governing the Activities of States in the Exploration and Use of Outer Space" in 1963 (Wolter 2003, p. 4) The Declaration has since been supplemented by three resolutions laying down the legal principles applicable to the exploration and exploitation of outer space, a "Declaration on International Cooperation in the Exploration and Use of Outer Space for the Benefit and in the Interest of All States, Taking into Particular Account the Needs of Developing Countries," and five treaties and agreements governing the use of space and space-related activities. (United Nations Treaties and Principles on Space Law) These treaties, agreements and principles are collectively known as the "United Nations Treaties and Principles in Outer Space." Nuclear weapons are forbidden, but other weapons (kinetic weapons, lasers) are permitted. (Although nuclear weapons are banned, it is recognized that some uses of nuclear power are needed in space, the Treaties and Principles provide for safety in its use, mitigation of risks, and liability for states that fail to control the nuclear power or its sources.)

The very nature of outer space is such that spacecraft do not have the same ability to control their flight paths that aircraft operating within the atmosphere have, (Braeunig 1997-2008) and the cost of a space program that could interdict is large, (Fox 2007) so a rule requiring interdiction of belligerents' weapons in space by the neutral does not make sense. Spacecraft and satellites in orbit pass above both belligerents and neutrals and cannot avoid doing so, being subject to the laws of celestial mechanics. Accordingly, the notions of territorial control that apply in the laws of the sea and the regulation of aircraft, cannot apply in outer space. If neutrals were required to exercise control over the use of outer space in the same way they exercise control over air traffic in the skies above their territories, it would be practically impossible to maintain neutrality at all.

Similarly, recognizing the impossibility of neutrals interdicting belligerent Internet use of the neutral's information infrastructure without prohibitive costs or unacceptable consequences for the neutral's licit use of its own infrastructure: "a state may not be able to prevent [cyber] attacks from leaving its jurisdiction unless it severs all connections with computer systems in other states." (Brown 2006, p. 210) This indicates that the appropriate rule for Internet use is more like the rule for space than the rule for air or land traffic, even when the use involves cyber weapons or information useful to the belligerent for military purposes (telemetry, GPS, weather data, etc.). Such acceptable use would, of course, apply to all belligerents, because the rules of neutrality prohibit the neutral state favoring one side in any way over the other side. (Brown 2006, p. 211)

5. Conclusion

Phillip Jessup, in 1936, concluded, "There is nothing new about revising neutrality; it has undergone an almost constant process of revision in detail." (Jessup 1935-6, p. 156. Cited in Walker 2000, p. 109) With the advent of cyberwar, rules governing neutrality during periods of belligerency need to be reconsidered and revised yet again. The realities of the Internet age mean that weapons as well as information can move across communications networks in ways that were not possible or foreseeable during the earlier evolution of the laws of war and neutrality. Yet the paths that those weapons will take as they traverse the Internet on the way to their intended targets are beyond the knowledge or control of the belligerents that launch them. Detection, identification and interdiction by neutrals across whose territories the weapons may pass are impractical without sacrificing the utility of the networks for licit use by the neutrals and others, hence impossible.

However, it is the only the details of the rules of neutrality that must change. Neutrals will not be required to do what they cannot do, and will not be subject to attack when they do not detect, identify and interdict the flow of weapons through their information infrastructures. The key principle of neutrality requiring that neutrals do not knowingly and willingly participate in the belligerency, or favor one side over the other, can and must be retained.

Disclaimer: Opinions expressed in this paper are those of the authors and do not represent positions of George Washington University, or of the Information Resources Management College, the National Defense University, the Department of Defense, or the United States Government.

References

The Avalon Project: Documents in Law, History and Diplomacy. Yale Law School, Lillian Goldman Law Library. http://avalon.law.yale.edu/default.asp.

Boot, Max (2006) War Made New: Technology, Warfare, and the Course of History, 1500 to Today. New York: Gotham Books.

Braeunig, Robert A. (1997-2008) Orbital Mechanics. http://www.braeunig.us/space/orbmech.htm.

Brown, Davis, A Proposal for an International Convention To Regulate the Use of Information Systems in Armed Conflict, 47 Harv. Int'l L.J. 179 (2006).

Clark, A. C. (1907) Q. Asconii Pediani Orationum Ciceronis Quinque Enarratio. http://www.attalus.org/latin/asconius2.html#Milo.

Dupuy, Trevor N. et al. eds. (2003) Dictionary of Military Terms, 2nd Ed. New York: H.W. Wilson.

Elsea, Jennifer K. & Grimmett, Richard F. (2007) Declarations of War and Authorizations for the Use of Military Force: Historical Background and Legal Implications. Washington, DC: Congretional Research Service RL31133. http://www.fas.org/sgp/crs/natsec/RL31133.pdf.

Fox, Bernard et al. (2007) Guidelines and Metrics for Assessing Space System Cost Estimates. Santa Monica, CA: Rand Corporation. http://www.rand.org/pubs/technical_reports/2008/RAND_TR418.pdf.

The Gale Group, Inc. (2008) West's Encyclopedia of American Law, Edition 2. Farmington Hills, MI: Thomson Gale. http://legal-dictionary.thefreedictionary.com/neutrality.

Gat, Azar (2006) War in Human Civilization. Oxford: Oxford University Press.

Giles, Lionel (1910) Sun Tzu on the Art of War. http://www.chinapage.com/sunzi-e.html.

Grotius, Hugo (1925) Du Jure Belli ac Pacis [Of the Law of War and Peace] Libri Tres. Oxford: Clarendon Press. [Reproduced as a Special Edition (1984) Birmiingham, AL: Legal Classics Library.] In particular, see Chapter XVII: On Those Who Are of Neither Side in War.

Hall, W. E. (1874) The Rights and Duties of Neutrals, Longman's Green, London.

Hague Convention (V) respecting the Rights and Duties of Neutral Powers and Persons in Case of War on Land. The Hague, 18 October 1907. http://www.icrc.org/ihl.nsf/FULL/200?OpenDocument.

Hensel, Howard M. (2008) Legitimate Use of Military Force. Surrey, UK:Ashgate Publishing Group.

Hyde, Charles C. (1945) International Law Chiefly as Interpreted and Applied by the

United States, Vol. 3. New York: Hachette Book Group USA (Little Brown & Co.).

International Humanitarian Law - Treaties & Documents by Date. International Committee of the Red Cross. http://www.icrc.org/ihl.nsf/INTRO?OpenView.

Jessup, Phillip and Deák, Francis (1935-6) Neutrality, Its History, Economics and Law: Vol. IV Today and Tomorrow. New York: Columbia University Press.

Johnson, Phillip A., et al. (May, 1999) An Assessment of International Legal Issues in Information Operations. Washington, DC: Department of Defense Office of General Counsel.

Kastenberg, Jushua E. (2009) "Non-Intervention and Neutrality in Cyberspace: An Emerging Principle in the National Practice of International Law." 64 A.F. L. Rev. 43.

Keegan, John (1993) A History of Warfare. New York: Alfred A. Knopf.

Kelsey, Jeffrey T. G. (2008) "Hacking into International Humanitarian Law: The Principles of Distinction and Neutrality in the Age of Cyber Warfare." 106 Mich. L. Rev. 1427.

Lauterpacht, Hersch, Oppenheim's International Law (7th Ed., 1948) London: Longmans, Green & Co.

Kolb, Robert (1997) "Origin of the twin terms jus ad bellum/jus in bello," International Review of the Red Cross, No. 320, p.553-562. Online at http://www.icrc.org/web/eng/siteeng0.nsf/iwplist163/d9dad4ee8533daef c1256b66005affef.

Lail, Benjamin (2002) Broadband Network and Device Security. Sydney: McGraw-Hill. http://books.mcgraw-hill.com/downloads/products/0072194243/0072194243_ch01.pdf.

Neff, Stephen C. (2000) The Rights and Duties of Neutrals. Manchester, UK: Manchester University Press.

Neutrality. (2008) West's Encyclopedia of American Law, Edition 2. http://legal-dictionary.thefreedictionary.com/neutrality.

Osmanczyk, Edmund Jan & Mango, Anthony (2004) Encyclopedia of the United Nations and International Agreements. Florence, Kentucky: Routledge.

Roberts, Adam and Guelff, Richard (1982) Documents on the Laws of War, 3d Ed. Oxford: Oxford University press.

"Rumsfeld Memo 16 October 2003" (2008) SourceWarch.
http://www.sourcewatch.org/index.php?title=Rumsfeld_Memo_16_Octo
ber_2003

Sullivan, Bob (2010) "Could Cyber Skirmish Lead U. S. to War?"
http://redtape.msnbc.com/2010/06/imagine-this-scenario-estonia-a-
nato-member-is-cut-off-from-the-internet-by-cyber-attackers-who-
besiege-the-countrys-bandw.html

"Testimony of OMB Director Nussle" (2008) The White House.
http://www.whitehouse.gov/omb/legislative_testimony_director_nussle_
021308

Tikk, Eneken et al. (2010) International Cyber Incidents: Legal Considerations. Tallinn: Cooperative Cyber defence Center of Excellence.

Tyson, Jeff. (April 3, 2001) "How Internet Infrastructure Works"
HowStuffWorks.com.
http://computer.howstuffworks.com/internet/basics/internet-
infrastructure.htm

United Nations Convention on the Law of the Sea (UNCLOS), (1982)
http://www.un.org/Depts/los/convention_agreements/convention_overvi
ew_convention.htm.

United Nations Convention on Prohibitions or Restrictions on the Use of Certain Conventional Weapons Which May Be Deemed to Be Excessively
Injurious or to Have Indiscriminate Effects, Protocol IV (1980).
http://www.un.org/millennium/law/xxvi-18-19.htm.

United Nations Treaties and Principles on Space Law (2010)
http://www.unoosa.org/oosa/en/SpaceLaw/treaties.html

von Glahn, Gerhard (1992) Law Among Nations: An Introduction to Public
International Law (6th ed.) New York: Macmillan.

Walker, George K. (November, 2000) "Information Warfare and Neutrality."
33 Vand. J. Transnat'l L. 1079.

"What is a packet?" (December 1, 2000) HowStuffWorks.com.
http://computer.howstuffworks.com/question525.htm

White, Matthew (2005) Source List and Detailed Death Tolls for the Twenti-
eth Century Hemoclysm. http://users.erols.com/mwhite28/warstat1.htm.

Wingfield, Thomas C. (2006) "When is a Cyberattack an 'Armed Attack?'
Legal Thresholds for Distinguishing Military Activities in Cyberspace."
Cyber Conflict Studies Association.
http://www.docstoc.com/docs/445063/when-is-a-cyberconflict-an-
armed-conflict

Wolter, Detlev (2003) Common Security in Outer Space and International
Law: A European Perspective. (Geneva: United Nations, UNI-
DIR/2005/29, 2006)

Changing the Face of Cyber Warfare with International Cyber Defense Collaboration

Marthie Grobler[1], Joey Jansen van Vuuren[1] and Jannie Zaaiman[2]
[1]Council for Scientific and Industrial Research, Pretoria, South Africa
[2]University of Venda, South Africa

Originally Published in the Conference Proceedings of ICIW 2011

Editorial Commentary

In South Africa, cyber security has been identified as a critical component contributing towards National Security. More rural communities are becoming integrated into the global village due to increased hardware and software corporate donations, the proliferation of mobile Internet devices and government programs aimed at bridging the digital divide through major broadband expansion projects. Comprehensive research conducted by the authors show that many of the new Internet users are not aptly trained to protect themselves against online threats, leaving them vulnerable to online exploits. This paper works towards the identification of any correlation between the economic development and mobile use propensity of Internet users with regard to National Security. The classification is based on availability of digital amenities, availability of and access to the Internet, the number of users per computer and the level of computer maintenance. Separate from these criteria, the availability of and access to the Internet via mobile phones has also been taken into consideration. The paper uses the results from the surveys to identify direct and indirect links between the factors in question. These results are then used to extrapolate the potential threat factor to National Security based on South Africans' cyber security awareness levels.

Abstract: The international scope of the internet and global reach of technological usage requires the South African legislative system to address issues related to the application and implementation of international legislation. However, legislation in cyberspace is rather complex since the technological revolution and dynamic technological innovations are often not well suited to any legal system. A further complication is the lack of comprehensive international cyber defense cooperation treaties. The result is that many countries are not properly prepared, nor adequately protected by legislation, in the event of a cyber attack on a national level. This article will address the international cyber defense collaboration problem by looking at the impact of technological revolution on warfare. Thereafter, the article will evaluate the South African legal system with regard to international cyber defense collaboration. It will also look at the influence of cyber defense on the international position of the Government, as well as cyber security and cyber warfare acts and the command and control aspects thereof. The research presented is largely theoretical in nature, focusing on recent events in the public international domain.

Keywords: collaboration, cyber defense, legislation, government responsibility

1. Introduction

The international scope of the internet and global reach of technological usage requires the South African legislative system to address issues related to the application and implementation of international legislation. However, the complexities of cyberspace and the dynamic nature of technology innovations requires a cyber defense framework that is not well suited to any current legal system. A further complication is the lack of comprehensive international cyber defense cooperation treaties, resulting in many countries not being properly prepared, or adequately protected by legislation, in the event of a cyber attack on a national level.

For the purpose of this article, cyber warfare is defined as the use of exploits in cyber space as a way to intentionally cause harm to people, assets or economies (Owen 2008). It can further be defined as the use and management of information in pursuit of a competitive advantage over an opponent, involving "the collection of tactical information, assurance that one's own information is valid, spreading of propaganda or disinformation among the enemy, undermining the quality of opposing force information and denial of service or of information collection opportunities to opposing forces" (Williams & Arreymbi 2007).

The article will address some of the aspects related to changing the face of cyber warfare, focusing specifically on international cyber defense collaboration. It will look at some international technological revolutions that had an impact on the international legal scope and briefly evaluate the South African legal system with regard to international cyber defense collaboration. The article will also address international cyber warfare and the influence of cyber defense on the international position of the Government. The article will conclude with recommendations on working towards international cyber defense collaboration.

2. Technological revolutions' impact on warfare

Modern society created both a direct and indirect dependence on information technology, with a strong reliance on immediacy, access and connections (Williams & Arreymbi 2007). As a result, a compromise of the confidentiality, availability or integrity of the technological systems could have dramatic consequences regardless of whether it is the temporary interruption of connectivity, or a longer-term disruption caused by a cyber attack (Warren 2008).

Battlespace, as implied by military use and warfare, is becoming increasingly difficult to define since advances in technology revolutionized the act of war. "Today, cyber attacks can target political leadership, military systems, and average citizens anywhere in the world, during peacetime or war, with the added benefit of attacker anonymity. The nature of a national security threat has not changed, but the Internet has provided a new delivery mechanism that can increase the speed, diffusion, and power of an attack." (Geers ND). Although the physical destruction of the internet infrastructure as a result of cyber warfare is unlikely, a number of technological exploits can be employed as part of a cyber warfare attack aimed at financial loss. These exploits include:

Probes - an attempt to gain access to a system;
Scans - many probes done using an automated tool;
Account compromise - hacking, or the unauthorized use of a computer account;
Root compromise - compromise of an account with system administration privileges;
Packet sniffing - capturing data from information as it travels over a network;
Denial of service (DoS) attacks - deliberate consuming of system resources to deny; and
Malicious programs and malware - hidden programs that causes unexpected, undesired results on a system (Owen 2008).

Technological revolutions in computers and electronics make major advances in weapons and warfare possible. It also extends to areas such as information processing and networks, communications, robotics and advanced munitions (O'Hanlon 2000). Technological revolutions enable countries to prepare offensive and defensive strategies in cyber space.

3. Evaluating the South African legal system with regard to international cyber defense collaboration

From recent activity, it is clear that both the South African Government, the defense environment and the business environment are becoming increasingly aware of the threats and implications enabled by the use of the cyber environment. It is also clear that the threats are becoming more sophisticated and advanced when used as an element of cyber warfare and cyber crime.

The internet is increasingly becoming more volatile and insecure. In fact, cyber terrorists have the capability to shut down South Africa's power, disrupt financial transactions, and commit crimes to finance their physical operations. Organized crime is also increasingly making use of the internet as a means of communication and financial gain. Therefore, South Africa needs a national cyber defense system to which everybody must obey.

3.1. The South African legal system

Over the past decade, South Africa has taken the first steps to protect its information. It has passed legislation starting with the South African Constitution of 1996, which protects privacy, and the ECT (Electronic Communications and Transactions) Act of 2002, which provides for the facilitation and regulation of electronic communications and transactions (ECT 2002).

In 2000, the PAIA (Promotion of Access to Information Act) No 2 as amended, was passed to give effect to Section 32 of the Constitution, subject to justifiable limitations (PAIA Act 2000). These limitations are aimed at the reasonable protection of privacy, commercial confidentiality and good governance in a manner that balances the right of access to information with any other rights, including the rights in the Bill of Rights in Chapter 2 of the Constitution (SA Constitution 1996). Linked to this Act is the PAIA Reg 187 Regulations regarding the promotion of information of access to information (Government Gazette 2003).

In 2002, the RIC (Regulation of Interception of Communications and Provision of Communication-related information) Act was passed to regulate the interception of certain communications, the monitoring of certain signals and radio frequency spectrums and the provision of certain communication-related information. This Act also regulates the making of applications for, and the issuing of, directions authorizing the interception of communications and the provision of communication-related information under certain circumstances (RIC Act 2002).

Towards the end of 2009, the South African Government passed two bills, namely the:

PPI (Protection of Personal Information) Bill that introduces brand new legislation to ensure that the personal information of individuals is protected, regardless of whether it is processed by public or private bodies (Giles 2010).

Information Bill that is meant to replace an existing piece of legislation, the Protection of Information Act of 1982. It deals with the protection of State information and empowers the government to classify certain information in order to protect the national interest from suspected espionage and other hostile activities (Republic of South Africa 2010).

Playing an important role in the South African legal system is international standards. ISO/IEC 27002 is an information security standard published by the International Organization for Standardization (ISO) and the International Electrotechnical Commission (IEC), originally published as ISO/IEC 17799:2005. It is entitled Information technology - Security techniques - Code of practice for information security management. This standard has been accepted by and adopted in South Africa (International Standards Organization 2008).

South Africa has also adopted the Council of Europe Cyber Crime Treaty in Budapest in 2001 but has not ratified it yet. The treaty contains important provisions to assist law enforcement in their fight against transborder cyber crime. Therefore, it is imperative that South Africa ratifies the cyber crime treaty to avoid becoming an easy target for international cyber crime. The ratification will hopefully be done soon, although the South African government seems to be presently focused on basic service delivery and more traditional crimes given the current local crime situation. However, steps to establish the Computer Security Incident Response Team (CSIRT) indicate that the aim to tackle cybercrime is gathering momentum.

3.2. The South African position on international cyber defense collaboration

In February 2010, South Africa published a draft Cyber security policy that would set a framework for the creation of relevant structures, boost international cooperation, build national capacity and promote compliance with appropriate cyber crime standards. Over the last five years, South Africa focused on modernizing and expanding information technology equipment, applications, and centralized hosting capabilities and network infrastructure. This was done as part of its strategy to fully modernize and integrate the national criminal justice system to the maximum benefit of society and at minimum cost to crime prevention agencies. This policy has not been adopted, but provides a first step from South Africa towards international cyber defense collaboration.

During a more recent attempt to international cyber defense collaboration, South Africa participated in the 12th United Nations Congress on Crime Prevention and Criminal Justice in Salvador, Brazil during April 2010. During this congress, delegates considered the best possible responses to cyber crime as the Congress Committee took up the dark side of advances in Information Technology. While advances in information technology held many benefits for society, its dark underside (computer-based fraud and forgery, illegal interception of private communications, interference with data and misuse of electronic devices) requires States to develop an organized, international response. Speakers at the congress remained undecided about the nature of the required response, with supporters of the Council of Europe's Budapest Convention on crime suggesting an expansion of the treaty, and others suggestion new multilateral negotiations (UN Information Officer 2010).

In general, governments are having a tough time keeping pace, and their responses to cyber crime is sadly lacking. In many countries, cyber crime damage economies and State credibility and further impedes national development. Cooperation in stamping cyber crime and protecting countries against cyber warfare is vital at all levels of defense, law enforcement, the judiciary and the private sector.

According to Markoff (2010), a group of cyber security specialists and diplomats, representing 15 countries (including South Africa) has agreed on a set of recommendations to the United Nations' Secretary General for negotiations on an international computer security treaty. In recent years, an explosion in cyber crime has been accompanied by an arms race in cyber weapons, as dozens of nations have begun to view computer networks as arenas for espionage and warfare. The recommendations to the United Nations from the specialists and diplomats reflect an effort to find ways to address the dangers of the anonymous nature of the Internet, as in the case of the object of a cyber attack misconstruing the identity of the attacker. Among the troubling issues is the existence of proxies. The report also suggests that "the same laws that apply to the use of kinetic weapons should apply to state behavior in cyber space." (Markoff 2010). The report recommends five steps to improve international cyber cooperation and security:

- Having more discussions about the ways different nations view and protect their computer networks, including the Internet;
- Discussing the use of computer and communications technologies during warfare;
- Sharing national approaches on legislation about computer security;
- Finding ways to improve the Internet capacity of less developed countries; and

- Negotiating to establish common terminology to improve the communications about computer networks (Markoff 2010).

The signers of the report are major cyber powers and of other nations: the United States, Belarus, Brazil, Britain, China, Estonia, France, Germany, India, Israel, Italy, Qatar, Russia, South Africa and South Korea. From a legal perspective, a number of concerns can be identified, such as:

- Lack of collaboration between industry and the defense environment;
- Capacity of the legal fraternity to comprehend the complexity of the cyber environment and to deliver a verdict based on a thorough understanding of the facts;
- Collaboration between countries and the agreements on protocols;
- Lack of collaboration between State Departments on cyber warfare and cyber crime;
- Lack of collaboration between municipalities, districts, regions and provinces; and
- Lack of collaboration between urban and tribal authorities.

Networked computers now control everything, including bank accounts, stock exchanges, power grids, the defence, the justice system and government. Networked computers also control all health records and crucial personal data. From a single computer an entire nation can be brought down. The authors are of the opinion that a series of regional conferences with all stakeholders involved and sponsored by private sector should be conducted. Significant progress has been made in South Africa, but commitments are required to draft a comprehensive Charter for South Africa and its unique situation.

4. International cyber warfare

The North Atlantic Treaty Organization (NATO) is only just beginning to recognize that the Internet has become a new battleground that also requires a military strategy. To counter such threats, a group of NATO members established a cyber defense centre in Tallinn. The 30 staffers at the Cooperative Cyber Defense Centre of Excellence analyze emerging viruses and other threats and pass on alerts to sponsoring NATO governments. Experts on military, technology, law and science are wrestling with such questions as: what qualifies as a cyber attack on a NATO member, and so triggers the obligation of alliance members to rush to its defense; how can the alliance defend itself in cyber space? Answers to these questions are strikingly different: Washington creates new funds for cyber defenses; Estonia is aiming to create a nation of citizens alert and wise to online threats (NATO ND).

The choice of Estonia as the home to NATO's new cyber war brain trust is not accidental. In 2007, Estonia suddenly found itself in the midst of cyber attacks. The fact that this happened in Estonia, a proud digital society, was eye opening. Back in 2007, Estonia's minister of defense stated that the attacks cannot be treated as hooliganism, but as an attack against the State. Nevertheless, no troops crossed Estonia's borders, and there was nothing that could be regarded as a conventional conflict. The United States clearly wants to take a military strategy approach. Estonia, on the other hand, prefers to demilitarize the issue by educating citizens on how to identify risks and promote a culture of cyber security, starting with schoolchildren. The Estonians have the right idea. A society of savvy citizens is the best defense (Geers ND).

In response to the cyber attacks on Estonia in 2007 and Georgia, NATO set up a coordinated cyber defense policy with a quick-reaction cyber team on permanent standby. This, however, has not stopped the constant attack on NATO computers (Gardner 2009).

5. Influence of cyber defense on the international position of Governments

The opinion of international Department of Defense (DOD) officials is that cyber space is a domain available for warfare, similar to air, space, land, and sea (Wilson 2007). As a result, any cyber attacks can have either a direct or an indirect influence on the DOD. Accordingly, the DOD needs to consider the potential effects of an emerging military-technological revolution that will have profound effects on the way wars are fought. Growing evidence exists that over the next several decades, the military systems and operations will be superseded by new, far more capable means and methods of warfare by new or greatly modified military organizations (Krepinevich 2003).

The DOD views information itself as both a weapon and a target in warfare. In addition, it provides the ability to disseminate persuasive information rapidly in order to directly influence the decision making of diverse audiences. By incorporating the cyber domain in the cyber defense structure, a number of new aspects come into play that may have an influence on the manner in which the DOD reacts to cyber attacks:

- New national security policy issues;
- Consideration of psychological operations used to affect friendly nations or domestic audiences; and
- Possible accusations against the State of war crimes if offensive military computer operations or electronic warfare tools severely disrupt critical civilian computer systems, or the systems of non-combatant nations (Wilson 2007).

An example of the last bullet point: if wrongful acts are committed inside a country, the State can be held responsible for these acts, since the State is obliged to fulfill the interest of the entire international community. If a representative of a State organ or a private person acting on the State's behalf committed an act, the act may be attributed to the State (Article 3 ILC Draft Articles). The physical location of a computer or hardware used in a cyber attack does not (and should not) allow for attributing that cyber attack to a particular State. Such an assumption would be greatly unjustified, since a State does not carry the responsibility for actions of its residents operating hardware located within its territory.

The State, however, can be held responsible in the light of existing international law doctrine, for a breach of an international obligation. This obligation relates not to actions but to omissions, i.e. for not preventing that attack to take place. This interpretation is derived from the wording of Article 14(3) of the International Law Commission (ILC) Draft Articles, which provides that a State may be held responsible for the conduct of organs of an insurrectional movement, if such an attribution is legitimate under international law. The State has therefore an obligation to show best efforts, and to take all "reasonable and necessary" measures in order to prevent a given incident to happen. The occurrence of this obligation was best reflected in the International Court of Justice (ICJ) case concerning the United States diplomatic and consular staff in Teheran. In its decision, the ICJ found that the overriding of the United States embassy in Teheran does not free Iran from the responsibility for that incident, although it also cannot be attributed to Iran (Kulesza 2010).

The State is also responsible for providing sufficient international protection from cyber attacks conducted by its residents from its territory. It is the duty of any State from whose territory an internationally wrongful act is conducted to cooperate with the victim's State and to prevent future similar harmful deeds. If the State itself is not capable of protecting the interests of another sovereign, it may also not allow for private persons acting from within its territory to inflict damage or create danger to that the other State while they are protected by its immunity. Under such an interpretation, Russia's denial to persecute the perpetrators of the attack against Estonia would constitute an internationally wrongful act, while Israeli involvement and punishment of the actors behind the Solar Sunrise attack on United States Airforce databases using the Texas internet provider exonerates them from any international responsibility (Kulesza 2010).

In this light, it is therefore the obligation of the South African government to launch and support awareness projects to prevent these attacks from inside its borders. This also includes the establishment of a CSIRT, as proposed in the draft South African Cyber security policy. Currently, South Africa is one of only a handful of countries that does not have a running CSIRT, putting South Africa in a disadvantaged position with regard to cyber attack and defense (FIRST 2009).

6. Working towards international cyber defense collaboration

Cyber warfare is an emerging form of warfare not explicitly addressed by existing international law. While most agree that legal restrictions should apply to cyber warfare, the international community has yet to reach consensus on how international humanitarian law (IHL) applies to this new form of conflict (Kelsey 2008). In particular, there is a need for an international consensus on the due diligence criteria which have to be fulfilled by a State in order to avoid international responsibility for failing to protecting other sovereigns from cyber attacks conducted from its territory.

Another crucial issue would be to establish the standards for releasing a State from any international responsibility for not providing due diligence: would the adoption of specific provisions in national criminal laws be sufficient or would State authorities need to initiate a criminal investigation effectively? It should also be clarified whether a due diligence standard can be set post factum – after an attack had already taken place (Kulesza 2010). In South Africa, this is not possible.

A suggested approach to create Nation State responsibility in building a credible cyber system involves the following steps:

- Developing a national strategy and making sure all agencies and major stakeholders follow it;
- Establishing a national endorsement body for cyber security;
- National coordination mechanism;
- Inclusion of all professional communities and private sector, and others in national cyber security effort; and
- Providing necessary resources and institutional changes (Tiirmaa-Klaar 2010).

If all the States internationally can implement their own credible cyber system, cooperation on an international cyber defense level will be easier to realize. As an initial attempt to enable a more uniform cyber defense system, the European Commission is planning to impose harsher penalties for cyber crimes. Large-scale attacks in Estonia and Lithuania in recent years have highlighted the need for a stronger stance on cyber crime. Estonia, Lithuania, France and the United Kingdom also have longer sentences for such crime, and the European Commission is looking to harmonize practice across the member states. United States president Barack Obama has declared cyber crime to be a priority. In addition to stronger laws, the European Union is looking to set up a system through which member states can contact each other quickly to notify one another of attacks. That would help to build a picture of the scope of cyber crime (Geers ND).

7. Conclusion

The Internet has changed almost all aspects of human life, also including the nature of warfare. Every political and military conflict now has a cyber dimension, whose size and impact are difficult to predict. "The ubiquitous nature and amplifying power of the Internet mean that future victories in cyber space could translate into victories on the ground. National critical infrastructures, as they are increasingly connected to the Internet, will be natural targets during times of war. Therefore, nation-states will likely feel compelled to invest in cyber warfare as a means of defending their homeland and as a way to project national power" (Geers ND).

The international scope of the internet and wide reach of technological usage has a tremendous impact on the nature of war and crimes globally. This article showed the impact of technological revolutions on warfare, the South African legislative system affecting warfare and cyber war, and the need for international cyber defense collaboration.

References

ECT Act (Electronic Communications and Transactions Act No 25 of 2002). (2002). Available from: http://www.acts.co.za/ect_act/ (Accessed 10 October 2010).

FIRST. (2009). FIRST: Teams around the world. Available from: http://www.first.org/members/map/ (Accessed 14 October 2010).

Gardner, F. (2009). Nato's cyber defence warriors. BBC News. Available from: http://news.bbc.co.uk/ 2/hi/europe/7851292.stm (Accessed 22 September 2010).

Geers, K. (ND). Cyber Defence. Available from: http://www.vm.ee/?q=en/taxonomy/term/214 (Accessed 22 September 2010).

Giles, J. (2010). How will the PPI Bill affect you? Available from: http://www.michalsonsattorneys.com/ how-will-the-ppi-bill-affect-you/2586?gclid=COXtIKz6yKQCFcbD7QodHzHJDg (Accessed 10 October 2010).

Government Gazette. (2003). Vol. 451 Cape Town 15 January 2003 No. 24250. No. 54 of 2002: Promotion of Access to Information Amendment Act, 2002.

International Standards Organization. (2008). ISO/IEC 27005: 2005. Information security risk management. Available from: http://www.iso.org/iso/catalogue_detail?csnumber=50297 (Accessed 10 October 2010).

Kelsey, JTG. (2008). Hacking into International Humanitarian Law: The Principles of Distinction and Neutrality in the Age of Cyber Warfare. P1427. Available from: http://heinonline.org/HOL/Landing Page?collection=journals&handle=hein.journals/mlr106&div=64&id=&page= (Accessed 22 September 2010).

Krepinevich, AF. (2003). Keeping pace with the military-technological revolution. Available from: http://www.issues.org/19.4/updated/krepinevich.pdf (Accessed 22 September 2010).

Kulesza, J. (2010). State responsibility for acts of cyber-terrorism. 5th GigaNet symposium Vilnius, Lithuania.

Markoff, J. (2010). Step Taken to End Impasse Over Cybersecurity Talks. Available from: http://www. nytimes.com/2010/07/17/world/17cyber.html?_r=1 (Accessed 8 October 2010).

NATO. (ND). Defending against cyber attacks. Available from: http://www.nato.int/cps/en/natolive/ topics_49193.htm (Accessed 22 September 2010).

O'Hanlon, ME. (2000). Technological change and the future of warfare. Brookings Institution Press: Washington.

Owen, RS. (2008). Infrastructures of Cyber Warfare. Chapter V. In: Janczewski, L. & Colarik, AM. Cyber warfare and cyber terrorism. Information Science Reference: London.

PAIA Act (Promotion of Access to Information Act No 2 of 2000 as amended). (2000). Available from: http://www.dfa.gov.za/department/accessinfo_act.pdf (Accessed 10 October 2010).

Republic of South Africa. (2010). Protection of Personal Information Bill. Available from: http://www.justice.gov.za/legislation/bills/B9-2009_ProtectionOfPersonalInformation.pdf (Accessed 10 October 2010).

RIC Act (Regulation of Interception of Communications and Provision of Communication-related information Act. (2002). Available from: http://www.acts.co.za/ric_act/whnjs.htm. (Accessed 10 October 2010).

SA Constitution. (1996). Available from: http://www.info.gov.za/documents/constitution/index.htm (Accessed 10 October 2010).

Tiirmaa-Klaar, H. (2010). International Cooperation in Cyber Security: Actors, Levels and Challenges. Cyber security 2010, Brussels, 22 September 2010 (Conference).

UN Information Officer. (2010). Delegates Consider Best Response to Cybercrime as Congress Committee - Takes Up Dark Side of Advances in Information Technology. Available from: http://www.un.org/News/Press/docs/2010/soccp349.doc.htm (Accessed 10 October 2010).

Warren, MJ. (2008). Terrorism and the internet. Chapter VI. In: Janczewski, L. & Colarik, AM. Cyber warfare and cyber terrorism. Information Science Reference: London.

Williams, G. & Arreymbi, J. (2007). Is cyber tribalism winning online information warfare? ISSE/ SECURE 2007 Securing Electronic Business Processes (2007): 65-72, January 01, 2007.

Wilson, C. (2007). Information Operations, Electronic Warfare and Cyberwar: Capabilities and related policy issues. CRS report for congress. Available from: www.fas.org/sgp/crs/natsec/ RL31787.pdf (Accessed 17 September 2010).

An Exceptional War that Ended in Victory for Estonia, or an Ordinary e-Disturbance?
Estonian Narratives of the Cyber-Attacks in 2007

Kari Alenius
Department of History, University of Oulu, Finland
Originally Published in the Conference Proceedings of ECIW 2012

Editorial Commentary

The paper discusses the Estonian Cyber attacks. In the spring of 2007 Estonia became the victim of a large-scale cyber-attack. Estimates of the significance of these events vary both in and outside of Estonia. For those who regard the events as being exceptionally important, the cyber-attacks launched against Estonia are seen as a milestone of modern warfare. Sometimes the term "Web War One" has even been used. At the other extreme, the events have been underestimated and their distinctiveness has been disputed. This study does not attempt to answer the question of which perspective is "right" and which is "wrong", especially when it is particularly difficult to provide an objective answer to this type of question. Instead, this study analyses Estonian interpretations of what occurred. The central elements of the Estonian main narrative crystallized during the summer and fall 2007. The narrative came to be composed of a few key elements describing the entire conflict in general and in a stereotypical way.

Abstract: In the spring of 2007 Estonia became the victim of a large-scale cyber-attack. Estimates of the significance of these events vary both in and outside of Estonia. For those who regard the events as being exceptionally important, the cyber-attacks launched against Estonia are seen as a milestone of modern warfare. Sometimes the term "Web War One" has even been used. At the other extreme, the events have been underestimated and their distinctiveness has been disputed. This study does not attempt to answer the question of which perspective is "right" and which is "wrong", especially when it is particularly difficult to provide an objective answer to this type of question. Instead, this study analyses Estonian interpretations of what occurred. The central elements of the Estonian main narrative crystallized during the summer and fall 2007. The narrative came to be composed of a few key elements describing the entire conflict in general and in a stereotypical way.

Keywords: rhetoric, narratives, cyber-attack, Estonia

1. Introduction

For this study, the internet resources of Estonia's leading media outlets *Eesti Päevaleht* (EPL), *Postimees* (PM), *Õhtuleht* (OL), *Eesti Rahvusringhääling* (ERR) have been examined from the end of April until the end of June. In addition, a Google search of published material on the internet has been done using keywords *cyber-attack, Estonia, 2007*, and their Estonian equivalents. In this way, individual published speeches have been found from among other publications and from the home pages of other quarters. In the case of the four aforementioned media outlets it is apparent that the Google search yielded almost exactly the same results as a systematic review of internet newspaper archives. Thus, it can be concluded that key Estonian internet data has been analyzed for this study.

The aim is to find out what kinds of narratives of these events were created in Estonia and why these narratives were a certain kind. As an alternative to narrative we can speak of discourses or mental images. Regardless of what the selected term is, in question is the examination of the processes that essentially guide human activity. Multiple sciences have convincingly demonstrated that above all, people act on the basis of their mental impressions, and not on the basis of "objective facts" that are empirically observable, although of course the former are built upon the latter. In many ways, mental images are stereotypical, in other words, simplified and coloured models of reality, and for the most part they arise as a result of largely unconscious and to a more minor degree, conscious psychological processes (Fält, 2002, 8-10; Ratz, 2007, 189-195).

Along with empirical findings, mental images are influenced by an individual's beliefs, fears, hopes, and all of the factors behind these – in short, the whole experiential history of an individual and their perception of the world. If a group of people have sufficiently similar images regarding a subject, then we can speak of collective images. Narratives and discourses partly reflect already existing mental impressions. They are also partly used for constructing images, clarifying images for oneself and spreading them to other people (Fält, 2002, 9-11; Ratz, 2007, 199-213). In any case, the importance of mental images in interaction between people and throughout the course of history justifies why, in the case of Estonia's cyber-attacks of 2007, it makes sense to analyze mental images and narratives created by events in Estonia.

2. The cyber-attacks in 2007 and their contexts

To understand the narratives generated there is first reason to briefly explain actual events and their associated contexts. The cyber-attacks were related to Estonia's so-called Bronze Soldier uproar. In 1947 the Soviet Union had set up a military statue in the center of Estonia's capital, Tallinn, the official name of which was "a monument to the liberators of Tallinn". After Estonian independence (1991) the fate of all Soviet monuments came into question. The Bronze Soldier was left in place but became a memorial for all those that had fallen during the WWII. However, these changes did not prevent the statue from becoming the focus of disputes. Some Estonian Russians organized celebrations annually near the statue on May 9 on Russia's so-called Victory Day, as well as on September 22, the anniversary of Tallinn's "liberation". In the minds of many

57

Estonians, these kinds of celebrations were hostile actions towards Estonia, as on the Estonian side the statue was often regarded as a symbol of Soviet occupation. From the Estonian perspective, the open show of Russian and Soviet symbols during the celebrations was a glorification of the occupation and a distortion of history (Kaasik, 2006, 1893-1916).

On May 9, 2006 there was a confrontation at the statue in which Russian celebrators attacked protesters carrying the Estonian flag. After the conflict, demands that the statue be removed from Tallinn's center and placed elsewhere strengthened. At the beginning of 2007 the Estonian parliament adopted two laws on the basis of which the Bronze Soldier and other similar monuments, as well as any dead buried in connection with them could be moved to a more suitable location. Preparations to move the Bronze Soldier and the Soviet soldiers buried nearby began on April 26, 2007. The statue and its surroundings were isolated with fences and unauthorized access to the site was prevented. The same evening, the Russians opposed to the operation were involved in large-scale rioting and sabotage in the center of Tallinn, and the unrest continued the following night. The Bronze Soldier was moved as planned to Tallinn's military cemetery and opened to the public on April 30, and the situation in Tallinn calmed down (RKK, 2007, 1-3).

At the same time as the riots, targeted cyber-attacks against Estonia began on April 27, which mainly targeted the websites of Estonian state institutions. The attacks mainly consisted of massive spamming and DDOS attacks. On the last day of April the extent and technical level of the attacks rose sharply and the main focus became the DNS system of Estonian servers. The number of sites expanded to include Estonian Internet service providers and the Estonian media. The attacks continued in varying intensity on a daily basis until mid-May, after which the situation almost returned to normal. Most of the cyber-attacks came from Russia, and based on the technical factors and large-scale resource requirements for the attacks, this suggests that the Russian government was involved in the attacks. The Russian state naturally denied involvement (RKK, 2007, 2-4; Saarlane, 2007-05-17).

Russia has also denied responsibility for other aggressive actions against Estonia. However, as early as the beginning of 2007, Russia's state leadership warned Estonia about moving the Bronze Soldier, and on April 23, left Estonia a formal diplomatic note concerning the issue (ERR, 2007-04-27). Already before the relocation of the statue there had been intensifying anti-Estonian verbal attacks in the state-controlled Russian media, and during the riots, the Russian Embassy in Tallinn, at the very least, kept close ties with the leaders of the riots. In Moscow, anti-Estonian protesters surrounded the Estonian Embassy for a week, apparently with the consent of the government, and prevented it from operating normally. In practice, Russia also undertook economic sanctions against Estonia and began a boycott of Estonian products in Russia (RKK, 2007, 3-4).

3. The main narrative

When we examine the reaction of the Estonian public to these cyber-attacks, it is apparent that quite soon after the onset of the attacks public debate began to develop in two rival narratives. On one hand, there was the mainstream public debate, which can be called the main narrative, and on the other hand, there was a side narrative that received less publicity, which can also be called a counter narrative. The main narrative was represented by Estonia's state leadership as well as the majority of journalists and IT professionals that publicly commented on the issue. The creators of the counter-narrative consisted of a few individual commentators who belonged to the two latter groups.

During the first three days of the cyber-attacks there was uncertainty and confusion among the Estonian public, which did not yet provide sufficient conditions for the birth of a narrative. In the first few days the main focus was on the riot and its aftermath, which is understandable as this had never been seen before in Estonia and in its drama, it had a high news value. Then and in the next couple of days, the cyber-attacks were also relatively few, and no clear information was available regarding their nature and origin. The issue was also new and unexpected: it was not anticipated, and there were no precedents in Estonia or elsewhere in the world on the basis of which an image could immediately be built. Thus, public uncertainty and confusion was apparent in that the media took a neutral tone in news regarding the matter. For example, these reported that the websites of Estonian state institutions had been attacked or were being harassed, but other evaluations of these events were not presented (ERR, 2007-04-28; ERR 2007-04-29).

The birth of a main narrative can be considered April 30, the date on which the first indicative commentaries appeared (OL, 2007-04-30; Virumaa Nädalaleht, 2007-04-30). Over the next two weeks, the mainstream image presented and the narrative of the incident broadened and took its essential form. During the second half of May, a few additional elements related to the end of the attacks were added to this. Then, there were more time and better conditions for drawing conclusions and forming an overall picture. However, the most active phase of public debate occurred in mid-May, and from the perspective of the media, the actuality of the topic began to wane after this. During the summer and fall of 2007 the topic was rarely returned to in the Estonian public, but on the other hand, the main narrative only took its final form during this time.

The declaration of Estonia's Justice Minister Rein Lang to Estonian television on the evening of April 30 acted as the initiator of the main narrative. Lang stated that investigations into the IP addresses of the attackers had revealed that the attacks originated in Russia, among others, from government institutions in Moscow (OL, 2007-04-30). During the next couple of days, one of the basic elements crystallized among the Estonian public discussion: Russia was the attacker (Delfi, 2007-05-01; OL, 2007-05-03). Although a few news reports stated that the majority of the attacks came from other addresses besides those under the direct control of the Russian government, and additionally the so-called botnet technique hampered clarification regarding the origin of the attacks, the guilty party was now known (ERR, 2007-05-04).

If the Russian state did not itself organize the attacks, it was responsible for them, as it could have chosen to prevent them. Additionally, "Russia" was guilty as, in any case, the attacks came from there, whether they were implemented privately or by the government (Delfi, 2007-05-01; OL, 2007-08-09). The conclusion that Russia was guilty was probably affirmed by other circumstantial evidence such as the earlier threats and verbal attacks against Estonia by Russian government leaders and the media, as well as Russia's suspected involvement in the rioting, at the very least as an instigator, and where necessary, as an advisor.

The identification of an enemy was a relief to Estonians, as afterwards, it was easier to interpret the situation and more possible to design countermeasures – at least at the level of beliefs. A vague and faceless enemy is always experienced as more fearsome (Zur 1991, 345-347). In question was a general human psychological reaction, which one commentator descriptively put into words at the beginning of May: "...The issue also has a good side. Events on the streets of Tallinn illustrate who our enemies are, and how many of them there are. There are no more illusions. Enemies cannot be integrated" (Delfi, 2007-05-01).

When an enemy had been found for the main narrative, the narrative could be begun and almost inevitably, one began to build using the logic and structure of the general image of the enemy. Since this was a new type of situation which did not fully recall traditional war (for instance, an official declaration of war and the conventional use of military force were lacking), all of the typical elements in the image of the enemy could not be used. Nevertheless, a few main elements were included in the Estonian main narrative. Firstly, the terminology used portrayed a war and an enemy. There was an enemy that attacked and one's own country, which repelled these attacks. A clear polarization between "us" and "others" is necessary in perceiving the existence of an enemy or another party (Zur, 1991, 345-346).

Secondly, the perception of the enemy is related to clear valuations of "good" and "bad". One's own side represents good and acts properly, while the other party represents bad and acts incorrectly, both on a theoretical–moral and practical level (Zur, 1991, 345-353). According to this polarization of values there was no understanding shown for the acts of the enemy in the Estonian public, but they were categorically condemned. Usually, there is no room for pondering the actions of the enemy in the sense that consideration would be given to why the enemy views the situation of conflict in a different way, and could the enemy have any "reasonable" or even "legitimate" grounds for its actions, from its perspective. In the mainstream Estonian public debate, those responsible for these cyber-attacks were explicitly in the wrong, malicious, and criminal (Delfi, 2007-05-01; ERR, 2007-05-05; Arvutikaitse, 2007-05-09).

The third characteristic adopted in the Estonian main narrative can be considered the typical manner in which the strengths and weaknesses of the enemy were brought to light. An appropriate balance between these two characteristics is always sought in portraying the enemy. The enemy must be sufficiently strong so that the threat to one's own side is taken seriously and that there is sufficient readiness to fight against the enemy, and if necessary, to make sacrifices in order to achieve victory. At the same time, victory over a strong enemy emphasizes the heroism and ability of one's own side and acts as a mental factor in strengthening the community. However, in emphasizing the strength of the enemy one should not go too far, as if it is portrayed as being too strong, this can result in hopelessness and defeatism among one's own community (Zur, 1991, 346-360).

In mainstream Estonian debate, the strength of the enemy was emphasized by explaining openly and in detail how wide-ranging and how many types of cyber-attacks had been made against Estonia. At the same time however, it was remembered to note that these attacks had been repelled. If in some cases the enemy had gained an advantage, then this advantage at least was temporary and limited. The counter-measures of one's own side had already gained control of the situation and no vital area had truly been in danger (ERR, 2007-05-01; OL, 2007-05-03). Thus, both the listeners and perhaps also the narrators of this narrative were able to feel safe in relation to the overall developments and final result of the war.

One could also feel safe in regard to the fact that in spite of its strength, in the case of Estonia, the enemy was weak and inferior, according to classic models. These characteristics can occur, for example, in the ridiculousness of the enemy. A comparison that is directly or apparently made to one's own side strengthens the opposing characteristics of one's own side. In Estonian mainstream public debate, this element was reflected in the good-natured comments of a few IT professionals regarding the simplicity of some cyber-attacks and the fact that they were easily deflected. The inability of the attackers to understand that their IP addresses were also easily found and that they revealed themselves was also wondered at publicly. No setbacks in these contexts were mentioned (ERR, 2007-05-04).

4. The competing side narrative

Within the competing side narrative, these aforementioned issues were mostly denied or put in perspective. The common basic premise was that Estonia had ended up in difficulties. The critique was not so great that it would have questioned belonging to the same community (Estonia/Estonians) or that the existence of the conflict would have been in dispute. However, the description of details and their interpretations differed.

This counter model of interpretation did not interpret the Russian state as the main opponent. This did not actually mean that it took a positive stance towards Russia, but it emphasized the difficulty of tracing the nature of the attacks as well as the role of individual Russian entities. In practice, the data available was the same as that which supporters of the main narrative had, but supporters of the counter narrative thought that it justified lesser conclusions: the Russian state was not merely responsible. At the same time the intermittent success of cyber-attacks and at least one's own temporary insufficiency to respond to them was also highlighted (EPL, 2005-05-17; Elamugrupp, 2007-06-30).

In the minds of those supporting a counter narrative, the question was also perhaps of a war, but according to their interpretation, the issue was not only about a simple series of successful defences, as was presented in the mainstream debate. At the most extreme, Estonian defenders of cyber-attacks were accused of overreacting and even unknowingly playing into the hand of the enemy: if the goal of the enemy was to isolate Estonia from the rest of the world, then the Estonians had ultimately done this themselves by blocking the access of foreign Internet addresses to Estonian websites (EPL, 2007-05-17).

There were relatively few statements that built on a side narrative in Estonian public debate, about one tenth of all the news and commentary. There was roughly the same number of statements belonging to a "gray zone", and it is difficult to classify these as belonging to either group. Therefore, about eighty percent of all the statements belonged to the group that built on the main narrative. The differences were not between different publications; for example, there were no significant differences between Estonia's leading Internet publications. Individual supporters of a side narrative could be found in different publications without that they would have been concentrated in any of them.

Possibly, these journalists were practicing a culture that was characteristic within conditions wherein which there was a free exchange of information, particularly of Western democracies, which a few researchers have referred to as a symbolic shadow-boxing. In conditions where there is freedom of information, it is considered the duty of the media to provide the general public with an image that is not too uniform, regardless of the issue and situation. If differing interpretations of the "facts" are not otherwise born, self-respecting journalists must create them if necessary, in the name of criticality and pluralism. This can lead directly to the aforementioned "shadow-boxing". In this case, the basic configuration of the crisis and the justification for defending one's own side is not put into question, but it is considered necessary to also search for mistakes and failures in one's own actions (Carruthers, 2000, 157-158).

5. Additional characteristics and further development

The key characteristics of the above-mentioned main and side narratives were created and established by mid-May. In the side narrative there was no apparent formation of additional characteristics after this, which is partly due to the fact that there were very few statements belonging to this side narrative after mid-May. Based on these few late comments, it is not possible to make any broader conclusions regarding possible changes in position (Vikerkaar 2008). In the case of the main narrative however, it is possible to continue an analysis of developments. By the second half of May two additional characteristics had joined the picture, and during the summer and fall of 2007 the image crystallized to take the shape of a few general interpretations.

The first additional characteristic was that by the end of May, it was already ventured to declare that Estonia was the victor in the war. No significant cyber-attacks had occurred for a week, and in perspective of the daily attacks that had occurred by the end of April and beginning of May, this seemed sufficient evidence to end the war. On the other hand, as logic regarding the image of the enemy entails, one's own side has no reason to lull themselves with a false sense of security. Once the enemy has been found he continues to be a potential enemy, and it is unrealistic to hope for a world without an enemy. In the main narrative it was remembered to emphasize that the attacks against Estonia and elsewhere were possible in future, even probable. For this reason, Estonians had to remain vigilant and develop their capacity to combat future attacks (OL, 2007-05-17; EPL, 2007-05-25).

The second additional feature was closely related to the previous one. In principle, it contained two conflicting sides of the same issue. On one hand it was reported that Estonia was an object of admiration for NATO allies, and the statements of allied representatives visiting Estonia were quoted frequently in public. Accordingly, the guests came to learn from the Estonians (Äripäev, 2007-05-18; PM, 2007-05-25). Praise received from others is always pleasant and helps to construct a positive image of oneself or of one's own group; to this extent, the quotation of statements had a clear general psychological background. On the other hand however, several Estonian statements which otherwise clearly belonged to the main narrative emphasized the need to gain support in rebuffing the cyber-attacks. Speeding up the construction of NATO's cyber defense center in Estonia was met with joy, and additionally, there was a desire for international reform regarding the definition of cyber-attacks. It was considered that the existing international agreements were outdated: they did not take the matter seriously enough, take into account the technological development in the field, nor allow for a sufficiently effective legal and practical response (PM, 2007-05-14; Virumaa, 2007-05-25).

It is clear that the achievement of international agreements and definitions that would obligate other countries to assist states that have become the target of cyber-attacks would have been to Estonia's advantage. For this reason, it was reasonable for Estonia's representatives to demand this in both Estonia's internal debate as well as abroad (EP, 2007-05-10). At the same time, assistance, in particular from other NATO countries, would have been welcome. In principle, however, it was questionable what other assistance would have been available from others if Estonia was already the world's leading expert in cyber defence. This paradox was not mentioned in Estonian statements. According to the principles of propagandistic communication, contradictory elements can be used in communication, as long as they are not presented at the same time (Zur, 1991, 350-351). Thus, in the Estonian main narrative these things – the Estonia in need of assistance and Estonia as the world's most skilled – never appeared in the same statements.

On the international scene, defining the cyber-attacks as a war was driven especially by the Estonian president Toomas Hendrik Ilves, as well as by Estonian members of the European Parliament (EP, 2007-05-10; President, 2007-06-18). For them, based on their positions and contacts, driving the national interests of Estonia abroad naturally fit well. At the same time, their positions were also heard by the Estonian public. The same message was forwarded particularly to the domestic audience by the speaker of the Estonian parliament Ene Ergma (EPL, 2007-05-25). When commander of the Estonian armed forces Ants Laaneots is added as a constructor of the main narrative, it can be said that Estonia's highest state leadership was more or less in favor of the main narrative. Out of all Estonia's influential public personas, Laaneots (PM, 2007-06-20), as well as former Prime Minister and Chairman of the leading right-wing party (IRL) Mart Laar (OL, 2007-08-09), most clearly stated that the Russian state was responsible for the cyber-attacks. Ilves and Ergma expressed the matter a little but more diplomatically, but even in their statements there was no doubt regarding the main opponent in the war.

6. Conclusions

Thus, the central elements of the main narrative crystallized during the summer and fall, so that over time the details and the exact course of events became side issues. These kinds of components fell or were dropped from the narrative and the image increasingly came to be composed of a few key elements describing the entire conflict in general and in a stereotypical way. All in all, it can be concluded that according to the Estonian main narrative, the cyber conflict consisted of the following components: 1) it was a war; 2) the Russian state was either directly or indirectly responsible for the attacks; 3) in question was a new, unprecedented kind of war; 4) the war ended in victory for Estonia.

Of the Estonian public debate, approximately eighty percent built on or supported the narrative described. In this sense, it can be regarded as a strong national narrative. The fact that in its contents it had a strong nationalistic emphasis and that its most visible supporters were individuals that belonged to a conservative right-wing also makes it a national narrative. As in many cases the same individuals also belonged to Estonia's state and political elite, the main narrative can also be described as Estonia's official narrative to a large extent. However, it was not a case of the Estonian government forcing members of the elite or the Estonian media to comply with this explanation. It was sufficient that the situation and conditions were such that the majority of those who participated in public debate came to similar conclusions on their own initiative. The experience of Estonian society coming under an unfair attack from the outside gave birth to a very large, uniform reaction of defence that was explained to oneself and to others in the form of this main narrative.

A side narrative was perhaps formed as a conscious counter-reaction to a main narrative that was experienced as being too uniform and thus propagandistic or implausible. It may also have been a case of differences in interpretation regarding other events, without any initial purpose of criticizing the mainstream narrative. The counter narrative questioned the nature of events as a war and preferred to support the interpretation of Internet harassment. At the same time, the interpretation that regarded the Russian state as opponent in the crisis was viewed as being too simplistic, and the private or unclear background of the attackers was referred to. The exceptionality of the events was also questioned and they were compared to known, large-scale operations of harassment and

Kari Alenius

damage initiated by private parties. Similarly, the fourth main characteristic of the main narrative, victory in the war, was not seen as a
legitimate interpretation: if there was no war, there was also no
victory. In addition, the success in combating the operations varied
according to this view.

In principle, both of these narratives were based on the same information regarding the events. Each narrative was also partly built
on the basis of general psychological models, in which the role of
apparent "facts" in shaping the narrative lessened. These narratives
used empirical construction materials, but their development also
partly followed models guided by a stereotypical, human way of
thinking. Thus, for instance, images of the enemy are generally similar to a great extent, regardless of the circumstances. The appropriate selection and appropriate interpretation of the information
available was essential so that they supported the perceived best,
simple enough explanation of the model – a narrative. It is not possible to determine the relative weight of conscious and unconscious
activity, but both have undoubtedly played a significant part in the
birth of these models. To uncover the specific characteristics of
these narratives, the content of the media in neighboring countries
(for instance, Baltic and Scandinavian countries) could be examined
for this same time period, but this requires a separate study in the
future.

References

Äripäev (2007-05-18) 'Eestist saab NATO kübersüda ja IT-polügoon',
    http://leht.aripaev.ee/?PublicationId=464dc490-fb94-4024-9b75-
    258ddc8543a9&articleid=12282&paperid=A4DE138A-6A0D-4C2A-
    A1B6-6613E673D67A
Arvutikaitse (2007-05-09), '9. maid tähistati küberrünnakuga',
    http://www.arvutikaitse.ee/9-maid-tahistati-kuberrunnakuga/
Carruthers, S. (2000) The Media at War: Communication and Conflict in the
    Twentieth Century, Basingstoke: Macmillan
Delfi (2007-05-01) 'Küberrünnak Eesti riigiasutustele',
    http://www.delfi.ee/archive/kuberrunnak-eesti-
    riigiasutustele.d?id=15733528

Case Studies in Information Warfare and Security

Elamugrupp (2007-06-30), 'Vaenlane kasutas kübersõjas müstilisi e-pomme',
http://www.elamugrupp.ee/modules.php?op=modload&name=News&file=article&sid=1067&mode=thread&order=0&thold=0

EP (2007-05-10), 'EP palub Euroopa Liidul näidata üles solidaarsust Eestiga', http://www.europarl.europa.eu/sides/getDoc.do?type=IM-PRESS&reference=20070507IPR06398&language=ET

EPL (2007-05-17) 'Sõja versioon 2.0 (beeta)',
http://www.epl.ee/news/arvamus/article.php?id=51087289

EPL (2007-05-25) 'Ergma: Eesti vastu suunatud küberrünnak ei jää EL-is viimaseks', http://www.epl.ee/news/eesti/ergma-eesti-vastu-suunatud-kuberrunnak-ei-jaa-el-is-viimaseks.d?id=51088437

ERR (2007-04-27) 'Venemaa andis Eesti suursaadikule pronksmehega seoses noodi', http://uudised.err.ee/index.php?0573764

ERR (2007-04-28) 'Valitsuse kommunikatsioonibüroo hoiatas valeteabe eest', http://uudised.err.ee/index.php?0573960

ERR (2007-04-29) 'Välismaised rünnakud häirivad valitsusasutuste veebilehti', http://uudised.err.ee/index.php?0574001

ERR (2007-05-01) 'Rünnakud Eesti küberruumi vastu on sagenenud',
http://uudised.err.ee/index.php?0574069

ERR (2007-05-04) 'IT-ekspert: Vene rünnakud serveritele on oskamatult tehtud', http://uudised.err.ee/index.php?0574202

ERR (2007-05-05) 'Politsei pidas kinni esimese küberrünnakus osaleja',
http://uudised.err.ee/index.php?0574259

Fält, O. (2002) 'Introduction', in Alenius K., Fält O. and Jalagin S. (eds.) Looking at the Other. Historical Study of images in theory and practice, Oulu: Oulu University Press.

Kaasik, P. (2006) 'Tallinnas Tõnismäel asuv punaarmeelaste ühishaud ja mälestusmärk', Akadeemia, no. 4.

OL (2007-04-30) 'Rein Lang: küberründed Venemaalt tulevad riiklikelt aadressidelt', http://www.ohtuleht.ee/227560

OL (2007-05-03) 'Venemaa küberrünnak Eesti pihta on Euroopa kohta tavatu', http://www.ohtuleht.ee/227851

OL (2007-05-17) 'Kübersõda karmistub', http://www.ohtuleht.ee/230007

OL (2007-08-09), 'Uurimise takistamine tõestab, et küberrünnak lähtus Venemaalt', http://www.ohtuleht.ee/241417

PM (2007-05-14) 'Aaviksoo rääkis NATO juhiga küberrünnakutest',
http://blog.postimees.ee/170507/esileht/siseuudised/260640.php

PM (2007-05-25) 'USA eksperdid: küberrünnak Eesti vastu äratas meid',
http://rooma.postimees.ee/040607/esileht/siseuudised/262679.php

PM (2007-06-20) 'Venemaa muutub Eestile üha ohtlikumaks',
http://rooma.postimees.ee/210607/esileht/siseuudised/267665.php

President (2007-06-18), 'Toomas Hendrik Ilves: "Kas küberrünnak on hädaolukord?',
http://www.president.ee/et/meediakajastus/intervjuud/3150-vabariigi-

Kari Alenius

president-ajalehele-frankfurter-allgemeine-zeitung-18-juunil-
2007/index.html
Ratz, D. (2007) 'The Study of Historical Images', Faravid, vol. 31.
RKK (2007) 'Moskva käsi Tallinna rahutustes', Rahvusvaheline Kaitseuuringute Keskus,
http://www.icds.ee/index.php?id=73&tx_ttnews%5Btt_news%5D=179&t
x_ttnews%5BbackPid%5D=99&cHash=a1145105e4
Saarlane (2007-05-17), 'Kreml eitas osalust Eesti küberrünnakutes',
http://www.saarlane.ee/uudised/uudis.asp?newsid=29727&kat=3
Vikerkaar (2008), 'Küberrünnakute moos aprillirahutuste kibedal pudrul',
http://www.vikerkaar.ee/?page=Arhiiv&a_act=article&a_number=4732
Virumaa (2007-05-25), 'VE: küberrünnakud',
http://www.virumaa.ee/2007/05/ve-kuberrunnakud/
Virumaa nädalaleht (2007-04-30), 'Minister Rein Lang: küberründed tulevad
Venemaa riiklikelt IP-aadressidelt',
http://www.vnl.ee/artikkel.php?id=6804
Zur, O. (1991), 'The love of hating: the psychology of enmity', History of
European Ideas, vol. 13, no. 4.

Estonia After the 2007 Cyber Attacks: Legal, Strategic and Organisational Changes in Cyber Security

Christian Czosseck, Rain Ottis and Anna-Maria Talihärm
Cooperative Cyber Defence Centre of Excellence, Tallinn, Estonia
Originally Published in the Conference Proceedings of ECIW 2011

Editorial Commentary

At the time of the state-wide cyber attacks in 2007, Estonia was one of the most developed nations in Europe regarding the ubiquitous use of information and communication technology (ICT) in all aspects of the society. Relaying on the Internet for conducting a wide range of business transactions was and still is common practice. Unlike other research on the Estonian incident, this case study shall not focus on the analysis of the events themselves. Instead it looks at Estonia's cyber security policy and subsequent changes made in response to the cyber attacks hitting Estonia in 2007. As such, the paper provides a comprehensive overview of the strategic, legal and organisational changes based on lessons learned by Estonia after the 2007 cyber attacks. The analysis of the paper provided herein is based on a review of national security governing strategies, changes in the Estonia's legal framework and organisations with direct impact on cyber security. The paper discusses six important lessons learned and manifested in actual changes: each followed by a set of cyber security policy recommendations appealing to national security analysts as well as nation states developing their own cyber security strategy.

Abstract: At the time of the state-wide cyber attacks in 2007, Estonia was one of the most developed nations in Europe regarding the ubiquitous use of information and communication technology (ICT) in all aspects of the society. Relaying on the Internet for conducting a wide range of business transactions was and still is common practice. Some of the relevant indicators include: 99% of all banking done via electronic means, over a hundred public e-services available and the first online parliamentary elections in the world. But naturally, the more a society depends on ICT, the more it becomes vulnerable to cyber attacks. Unlike other research on the Estonian incident, this case study shall not focus on the analysis of the events themselves. Instead it looks at Estonia's cyber security policy and subsequent changes made in response to the cyber attacks hitting Estonia in 2007. As such, the paper provides a comprehensive overview of the strategic, legal and organisational changes based on lessons learned by Estonia after the 2007 cyber attacks. The analysis provided herein is based on a review of national security governing strategies, changes in the Estonia's legal framework and organisations with direct impact on cyber security. The paper discusses six important lessons learned and manifested in actual changes: each followed by a set of cyber security policy recommendations appealing to national security analysts as well as nation states developing their own cyber security strategy.

Keywords: Estonia, cyber attacks, lessons learned, strategy, legal framework, organisational changes

1. Introduction

Over three weeks in the spring of 2007, Estonia was hit by a series of politically motivated cyber attacks. Web defacements carrying political messages targeted websites of political parties, and governmental and commercial organisations suffered from different forms of denial of service or distributed denial of service (DDoS) attacks. Among the targets were Estonian governmental agencies and services, schools, banks, Internet Service Providers (ISPs), as well as media channels and private web sites (Evron, 2008; Tikk, Kaska, & Vihul, 2010).

Estonian government's decision to move a Soviet memorial of the World War II from its previous location in central Tallinn to a military cemetery triggered street riots in Estonia, violence against the Estonian Ambassador in Moscow, indirect economic sanctions by Russia, as well as a campaign of politically motivated cyber attacks against Estonian (Ottis, 2008). By April 28th the cyber attacks against Estonia were officially recognized as being more than just random criminal acts (Kash, 2008). The details of the weeks that followed are described in (Tikk, Kaska, & Vihul, 2010).

The methods used in this incident were not really new. However, considering Estonia's small size and high reliance on information systems, the attacks posed a significant threat. Estonia did not consider the event as an armed attack and thus refrained from requesting NATO's support under Art. 5 of the NATO Treaty; instead, the attacks were simply regarded as individual cyber crimes (Nazario, 2007; Tikk, Kaska, & Vihul, 2010) or "hackitivism" as established by a well-known information security analyst Dorothy Denning (Denning, 2001). A further discussion on whether or not the 2007 attacks were an armed attack is beyond the scope of this paper. Many defence and security analysts have covered this particular topic and discussed e.g. the "juridical notion of information warfare" (Hyacinthe, 2009), a "taxonomies of lethal information technologies" (Hyacinthe & Fleurantin, 2007), formulated a "Proposal for an International Convention to Regulate the Use of Information Systems in Armed Conflict" (Brown, 2006), or "legal limitations of information warfare" (Ellis, 2006).

The incident quickly drew worldwide attention, and media labelled the attacks the first "Cyber War" (Landler & Markoff, 2007). This led to an overall "cyber war hype" that was continuously carried forward by media, researchers and policymakers. This exaggerating rhetoric was employed during following conflicts like Georgia 2008 or Kyrgyzstan 2009, and such misuse of terminology has already received a fair amount of criticism (Farivar, 2009).

The 2007 attacks have shown that cyber attacks are not limited to single institutions, but can evolve to a level threatening national security. Looking back, the Estonian state was not seriously affected since to a larger extent state functions and objects of critical information infrastructure were not interrupted or disturbed (Odrats, 2007). However, nation states did receive a wake-up call on the new threats emerging from cyber space, alongside with new types of opponents.

The following three sections will provide a comprehensive overview of major changes in Estonia's national cyber security landscape, namely the changes of national policy. As a result, several laws and regulations were introduced, while others were amended, and there were several changes in the organisational landscape.

This paper features six lessons learned that were identified as most remarkable in the case study of Estonia. It concludes with several strategic cyber security recommendations.

2. Development of national strategies

The benefits as well as threats of the use of Internet-related applications to information societies are identified by a number of Estonian high level policies and strategies.

The Estonian Information Society Strategy 2013 (MoEAC, 2006), in force since January 2007, promotes the broad use of ICT for the development of a knowledge-based society and economy. Given that cyber attacks on a scale matching that of Estonia in 2007 were unseen and likely unpredicted so far, it is not surprising that the risk of massive cyber attacks was not taken into serious consideration in the strategy – nor in other national policy documents from that era (see e.g. the implementation plan of the Information Society Strategy for 2007-2008, MoEAC, 2007)

The National Security Concept of Estonia published in 2004 (MoD, 2004) and the government's action plan in force at this time (Estonian Government, 2007) were no exception since these documents did not even mention possible cyber threats or related actions.

It was only after the 2007 cyber attacks that cyber security instantly found its way into the national security spotlight.

2.1. Policy and strategy responses since 2007

In July 2007, shortly following the cyber attacks, the Government approved the Action Plan to Fight Cyber-attacks (Kaska, Talihärm, & Tikk, 2010). In September 2007, the revised Implementation Plan 2007-2008 of the Estonian Information Society Strategy 2013 (MoEAC, 2007) was approved. The document holds a generic statement that critical information infrastructure should be developed in such a way that it operates smoothly in "emergency situations" (MoI 2009).

2.1.1. Cyber security strategy

In May 2008, the Estonian government adopted the newly drafted Cyber Security Strategy (CSS) as a comprehensive policy response to the cyber attacks. The strategy was prepared by a multi-stakeholder committee including relevant ministries, agencies and private sector representatives.

The CSS considers cyber security a national effort responding to the asymmetric threat posed by cyber attacks. The strategy underlines that state-wide cyber security requires active international cooperation and the promotion of global responses. On a national level, the strategy suggests implementing organisational, technical and legal changes. Further, it aims at developing an over-arching and sophisticated cyber security culture (MoD, 2008).

Based on a post-attack assessment of the situation in Estonia, the CSS identified five strategic objectives:

- The development and large-scale implementation of a system of security measures;
- Increasing competence in cyber security;

- Improvement of the legal framework for supporting cyber security;
- Bolstering international cooperation; and
- Raising awareness on cyber security.

In May 2009, the CSS implementation plan for the 2009-2011 cycle was adopted by the government. The plan called for concrete actions in five priority areas and became the main source for the comprehensive cyber security approach in Estonia (Estonian Government, 2009).

2.1.2. National Security Concept

The National Security Concept, which was updated and approved in May 2010, represents Estonian government's second major cyber security policy response. It recognizes Estonia's growing reliance on ICT along with the increasing threat posed by terrorists and organised crime groups. Cyber crime should receive special attention, and solutions are to be found in co-operation between agencies on both national and international level. Cyber security shall be ensured by "[...] reducing vulnerabilities of critical information systems and data communication connections". Critical systems shall stay operational, even if the connection to foreign countries is temporarily malfunctioning or has ceased to function. To support these actions, the necessary legislation should be developed and public awareness raised (MoD, 2010).

The National Security Concept led to the revised Guidelines for Development of Criminal Policy until 2018, published in October 2010. The Police shall focus on preventing the spread of malware and the growing number of "hacking" incidents. Furthermore "[t]he existence of a sufficient number of IT specialists in law enforcement agencies shall be ensured in order to set bounds to cyber crime more efficiently." (MoJ, 2010). Other strategies like the Estonian Information Society Strategy 2007-2013 have received only minor cyber security related amendments.

In addition, since the 2007 attacks, Estonia has become one of the major advocates of cyber security on the international level. As one result, NATO initiated the development of a unified strategy against cyber attacks (Blomfield, 2007) and in 2010 NATO adopted the new strategic concept that recognizes cyber attacks as a threat to the alliance and opts for the enhancement of alliance's and nations' capabilities to face the threat (NATO, 2010).

Moreover, Estonia has actively supported a number of international organisations such as the Council of Europe in its fight against cyber crime (MoFA, 2010a), Association of Southeast Asian Nations in promoting the harmonization of laws concerning cyber crime (MoFA, 2010b) and United Nations in contributing an expert to the task force on Developments in Information and Communication Technology in the Context of International Security (MoFA, 2010c).

2.1.3. Development in the legal field

The 2007 attacks prompted major changes in the Estonian legislative landscape and in some cases enhanced the changes already underway. Legal amendments involved several areas of law related to cyber security (see Table 1): criminal law (including aspects of criminal procedure) and crisis management law. The Estonian incident did not, however, directly touch upon the legal regime applicable to armed conflicts since the attacks were treated by national authorities as acts of crime.

Other laws such as the Electronic Communications Act were also updated but did not involve considerable changes in the context of cyber security (Estonian Government, 2010). Table 1.(Kaska, Tali-härm, & Tikk, 2010)

Table 1: Law related to cyber security

Constitutional law				
Fundamental rights and freedoms; Organisation of the state; Execution of public authority				
Private law	Public administrative law	Criminal law	Crisis management law	War-time law /national defence law
Information society services	General administrative procedure law supporting the accessibility of information society	Substantive criminal law	Critical infrastructure protection (CIP)	National defence organisation
eComms infrastructure provision	Availability of public information and public e-services	Criminal procedure law	Critical information infrastructure protection (CIIP)	National defence in peacetime
Provision of eComms services to end users	Data processing and data protection	International cooperation		National defence in conflict/wartime
General private law supporting the functioning of information society (eCommerce, digital signatures)				

3. Penal code

Mostly due to the need to harmonize the Estonian Penal Code with the Council of Europe Convention on Cyber Crime (Council of Europe, 2001) and the Council Framework Decision 2005/222/JHA of on attacks against information systems (Council of Europe, 2005) all cyber crime related provisions in the Penal Code were reviewed. The amendments targeted the provisions addressing attacks against computer systems and data, widened the scope of specific computer crime provisions (e.g. criminalizing the dissemination of spyware and malware), added a new offence of the preparation of cyber crimes, modified the provision concerning acts of terrorism and filled an important gap (Estonian Government, n d) in the Penal Code by enabling differentiation between cyber attacks against critical infrastructure (with the purpose of seriously interfering with or destroying the economic or social structure of the state) and ordinary computer crime (MoI, 2009).

3.1. Amendments relevant to criminal procedure law

The amendments in the Penal Code resulted partly from the regulatory limitations that arose in relation to the application of the Code of Criminal Procedure (CCP) to the 2007 attacks (MoJ, 2010b) as CCP §§ 110-112 maintain that evidence may be collected by surveillance activities in a criminal proceeding if the collection of evidence is a) precluded or especially complicated and b) the criminal offence under investigation is, at the minimum, an intentionally committed crime for which the law prescribes a punishment of at least three years' imprisonment (MoJ, 2010b). However, during the Estonian attacks in 2007 it became apparent that almost none of the committed offences met the threshold of "three years" imprisonment and that precluded the employment of surveillance measures (Estonian Government, 2007b). Therefore, the changes in the Penal Code prescribed higher maximum punishments and also corporate liability for cyber crime offences.

3.2. New Emergency Act

The new Emergency Act (EA) (MoI, 2009) was adopted in June 2009 and reviewed the current setup of national emergency preparedness and emergency management structure, including the responses to cyber threats.

Offering a comprehensive approach, the act foresees a system of measures which include preventing emergencies, preparing for emergencies, responding to emergencies and mitigating the consequences of emergencies ("crisis management") (MoI, n d). It is the providers of public services and information infrastructure owners that are tasked with everyday emergency prevention and ensuring the stable level of service continuity. Providers of vital services are obliged, among other assignments, to prepare and present a continuous operation risk assessment (EA § 38) and an operation plan (EA § 39) to notify the citizens about events significantly disturbing service continuity as well as to provide the necessary information to supervisory bodies. In addition to the above, there are certain provisions that specifically address threats against information systems, such as an obligation for the providers of vital services to guarantee the smooth application of security measures in information systems and information assets used for the provision of vital services.

4. Development of organisations

Before the 2007 cyber attacks Estonia had relatively few organisations dedicated to (national) cyber defence. Since then, Estonia has made some key organisational changes to better deal with the cyber threats. The most significant ones are described below.

A high level organisational change was the formation of the Cyber Security Council under the Government Security Committee, a body foreseen by the National Cyber Security Strategy. The Council reports directly to the Government Security Committee and is therefore well-placed for coordinating inter-agency and international cyber incident response.

4.1. EIC, CERT-EE and CIIP

Estonian Informatics Centre (EIC) is a state agency that is responsible for managing and developing public information services and systems (MoEAC, 2009). It is also tasked with providing cyber security for these services and systems. Even though a national CERT had been established in 2006 as a department of the EIC, its capabilities and experience were still quite modest at the time of the attacks. In 2009, as a result of the National Cyber Security Strategy, the Department of Critical Information Infrastructure Protection (CIIP) was added to the structure of EIC, in addition to the already existing CERT. The main tasks of the new department include supervising risk analyses of critical information infrastructures and developing protective measures.

4.2. Cyber defence league

During the cyber attack campaign, the Estonian CERT was assisted by an informal network of volunteer cyber security experts. This provided much needed additional capabilities, such as increased situational awareness, analysis capability, quick sharing of defensive techniques between targeted entities, as well as an extended network of direct contacts to international partners.

The roots of this informal group derive from the late 1990ies, when Estonia was adopting a national ID card system. Over the years, the network of professionals had also cooperated against criminally motivated cyber attacks targeting critical infrastructures (e.g., Estonian banks). A later development was the formalisation of this loose cooperation into the Cyber Defence League (CDL) in 2009. The Defence League is a volunteer national defence organization in the military chain of command. The CDL is part of the Defence League and unites cyber security specialists who are willing to contribute their time and skills for the protection of the high-tech way of life in Estonia, especially assisting the defence of critical information infrastructure. It is important to note that this is a defensive organisation, not designed to harass political adversaries in (anonymous) cyber attack campaigns. In January 2011, the CDL was reorganized into the Cyber Defence Unit of the Defence League, but the CDL name is still widely used.

CDL's key activities include organizing training and awareness events, as well as cyber defence exercises. In 2010, the CDL was involved with the Baltic Cyber Shield exercise organised by Cooperative Cyber Defence Centre of Excellence (Geers, 2010), the US-led International Cyber Defence Workshop, as well as a series of national exercises. The CDL is a good example of managing in a productive manner the expertise and enthusiasm of motivated cyber security specialists.

5. Six recommendations

Given that the major changes have been discussed above, the next section will feature six significant lessons learned from the 2007 cyber attacks against Estonia:

5.1. Comprehensive strategy approach

It is evident that Estonia has taken into account the lessons learned from the 2007 incident, the most significant step being the quick establishment of a comprehensive policy response which has led to the adoption and subsequent implementation of the national Cyber Security Strategy. The Estonian example emphasises the need for nation-wide cooperation and countermeasures against cyber crime, involving major stakeholders of the public and private sector.

It remains to be debated whether cyber security should be handled in a single comprehensive strategy or form a sub-section of all other relevant strategies touching upon ICT. However, considering the speed of technological advancements and comparing it with the speed of developing national strategies, the Estonian approach of having a single strategy might be the one more advisable.

The 2007 attacks triggered the cyber security strategy drafting in Estonia. However, countries should not wait for such triggers and should pro-actively conduct a thorough and comprehensive risk assessment of their cyber infrastructure. Furthermore, often only the context and additional information will reveal if the attack was launched with crime, espionage, terrorism or military motivation. Therefore, close cooperation between relevant agencies remains a sine qua non to success in this arena.

5.2. Politically motivated cyber attacks

Another aspect to consider is the shift of attention in terms of cyber security threats over the last decade. While the first half of the decade the cyber security focus was on criminal and espionage attacks (if recognised as a national security issue at all), the second half witnessed a surge in politically motivated cyber attacks (Nazario, 2009). The significance of this development is that targets have transformed. A politically motivated attacker is likely to attack visible and politically significant targets (such as the public website of a government agency or a company that has angered an interest group), which are of little interest to criminals and intelligence agencies. This shift in targets requires everyone to reassess their risks and security requirements.

Politically motivated actors can cover the entire spectrum of cyber attack, from high-profile strikes against critical infrastructure, to millions of pinprick attacks that can weaken the state over a long period of time (Lemay, Fernandeza, & Knight, 2010; Liles, 2010; Ottis, 2009). As the threat of politically motivated attacks threatening national security is not likely to go away in the foreseeable future, it must be addressed as a national security issue in order to get the full attention of policymakers.

5.3. Legal recommendations

An analysis of the Estonian legal order governing the domain of information society underlines that a secure information society needs to be comprehensively supported by norms involving several legal disciplines. The broad approach illustrated by the Estonian legal framework brings together the areas of private and public law, and completes the spectrum of cyber incident regulation by engaging criminal law, crisis management regulation and wartime law/national defence legal order. It is vital for countries to realize that the international cyber security regulation involves a wide range of legal areas and the review of relevant regulatory frameworks and the identification of possible uncovered "grey areas" is highly recommended.

Within national legal systems, a review of criminal law (penal law) appears to be a central issue. Attacks against critical (information) infrastructure, politically motivated cyber attacks, possible cases of cyber terrorism, as well as related provisions for investigation and prosecution, should all be reflected in the domestic criminal law or other national acts. Broad and inclusive national implementation of the Council of Europe Convention on Cybercrime is of crucial importance, especially considering the cross-border nature of cyber crime.

Additionally, the Estonian experience underlined the need to establish common security standards for all computer users, information systems and critical infrastructure companies (MoD, 2008). By 2011, steps have been taken to establish such standards for service providers within the framework of the Electronic Communications Act, but more detailed rules for end-users' conduct and/or legal obligations are still needed.

5.4. Exercises and education for the masses

A key component of enhancing (national) cyber security is cyber security awareness and education. This should not be limited to professionals in governmental or private institutions, but must cover the whole spectrum from a citizen using ICT for everyday things to senior policy makers, considering the skills and knowledge needed at every level. This includes law enforcement agencies and especially the judicial system that has a central role in interpreting the regulatory aspects of cyber security. By developing different solutions well suited for each groups, a broad and sophisticated cyber security culture can be implemented, as aimed for in the CSS.

Estonia recognized its lack of sufficient number of well-trained information security experts and developed a new Master's program for Cyber Security Studies in 2008. The Cyber Defence League is another venue for actively training experts in cyber security. Further measures, such as information campaigns for the secure use of the Internet, special classes in high school or vocational training should be considered by Estonia and other nation states.

Additionally, cyber security exercises organised both on national and international level serve as effective preparation to respond to cyber attacks. Exercises like Cyber Europe 2010 (ENISA, 2010) require efficient coordination between agencies and private shareholders and should be regularly conducted.

5.5. International relations

The attacks against Estonia in 2007 underlined the importance of international cooperation as it became even more apparent that in the context of responding to cyber threats, one country can do little alone. To that end, active participation in the work of major organizations dealing with cyber security requires keeping national developments and legal framework up to date and serves as a useful ground for new initiatives, further collaboration and regional or global forum. Moreover, the ratification of instruments such as the Council of Europe Convention of Cyber Crime that aim to harmonise cyber crime regulation worldwide should be supported and promoted.

Beside the political will for cooperation, national multi- and bilateral agreements, information sharing agreements, cooperation of law enforcement agencies, joint investigation teams, international exercises, formal and informal networks and other international initiatives are vital for effective prosecution and investigation of cyber crime offences.

5.6. Harnessing the volunteers

It is well known that most of the Internet infrastructure is owned and operated by the private sector. It follows that there is a pool of experts in the private sector, who could provide a meaningful contribution to national cyber security, regardless of their actual position in the private sector. This also includes experts in the public sector, who do not work in their area of expertise. Clearly, there are limits to the use of volunteers, whether their potential role is in offensive or defensive activities (Ottis, 2009). However, if proper legal, policy and operational frameworks are in place, volunteers can significantly increase national cyber security capability.

6. Conclusions

While in hindsight, the cyber attacks against Estonia were not as severe as often referred to, they still triggered an understanding of threats from cyber space as threats potentially affecting national security and prompted a wake-up call concerning the risks associated with the "careless use" of digital information technologies (e.g., Internet). For instance, the risk posed by politically motivated individuals should be regarded as a possible element of a serious threat to cyber security. By reviewing the strategic, legal and organisational changes that Estonia has undergone after the 2007 cyber attacks, this paper provides a concise list of key changes that have taken place on the legislative and administrative levels. While this paper describes some new assets that so far appear to be unique to Estonia, such as the formation of the Cyber Defence League, it offers several recommendations to national security planners performing beyond Estonia's national boundaries. Many of the aforementioned recommendations are not new; but they have passed a practical test through the real-life Estonian case study. Accordingly, these recommendations are more than a set of purely theoretical proposals. Lastly, based on the foregoing analysis, it is important to stress the fact that cyber security of a nation state can only be achieved by an interlocked approach covering national policies, its legal framework and organisations involving both public and private actors, as well as necessary changes identified by a realistic risk assessment.

Disclaimer

The opinions expressed here are those of the authors and should not be considered as the official policy of the Cooperative Cyber Defence Centre of Excellence or NATO.

Acknowledgement

We would like Mrs. Kadri Kaska and the unknown reviewer for their substantial comments they provided us with in the course of writing this paper.

References

Blomfield, A. (2007). Estonia calls for Nato cyber-terrorism strategy. Retrieved from http://www.telegraph.co.uk/news/worldnews/1551963/Estonia-calls-for-Nato-cyber-terrorism-strategy.html.

Brown, D. (2006) "A Proposal for an International Convention to Regulate the Use of Information Systems in Armed Conflict", Harvard International Law Journal, 47 (1), 179-221.

CDL. (n.d.). Cyber Defence League. Retrieved from http://www.kaitseliit.ee/index.php?op=body&cat_id=395.

Council of Europe. (2001). Convention on Cybercrime. Retrieved from http://conventions.coe.int/treaty/en/treaties/html/185.htm.

Council of Europe. (2005). Council Framework Decision 2005/222/JHA of 24 February 2005 on attacks against information systems. Official Journal L 69, 67-71.

Denning, D. E. (2001). Activism, hacktivism, and cyberterrorism: the internet as a tool for influencing foreign policy. Networks and netwars: The future of terror, crime, and militancy, 239–288.

Ellis, B. (2001) "The International Legal Implications and Limitations of Information Warfare: What Are Our Options?". Retrieved Mar. 2, 2011 from http://www.iwar.org.uk/law/resources/iwlaw/Ellis_B_W_01.pdf.

ENISA. (2010). EU Cyber Security Exercise 'Cyber Europe 2010'. Retrieved January 31, 2011, from http://www.enisa.europa.eu/media/press-releases/cyber-europe-20102019-cyber-security-exercise-with-320-2018incidents2019-successfully-concluded.

Estonian Government. (2007a). Programme of the Coalition for 2007-2011.

Estonian Government. (2007b). Explanatory Memorandum to the Draft Act on the Amendment of the Penal Code (116 SE) (In Estonian). Retrieved from http://www.riigikogu.ee/?page=pub_file&op=emsplain&content_type=a

pplica-
tion/msword&u=20090902161440&file_id=198499&file_name=KarS
seletuskiri
(167).doc&file_sise=66048&mnsensk=166+SE&etapp=03.12.2007&fd=
29.10.2008.
Estonian Government. (2009). Valitsus kiitis heaks küberjulgeoleku
strateegia rakendusplaani aastateks 2009–2011. Retrieved from
http://uudisvoog.postimees.ee/?DATE=20090514&ID=204872.
Estonian Government. (2010). Explanatory Memorandum to the Act amending the Electronic Communications Act (424 SE) (In Estonian). Retrieved from
http://www.riigikogu.ee/?page=pub_file&op=emsplain&content_type=a
pplication/msword&file_id=535868&file_name=elektroonilise side
muutmine seletuskiri
(424).doc&file_size=31650&mnsensk=424+SE&fd=.
Evron, G. (2008). Battling botnets and online mobs: Estonia's defense efforts
during the internet war. Georgetown Journal of International Affairs,
9(1), 121–126.
Farivar, C. (2009). A Brief Examination of Media Coverage of Cyberattacks
(2007 - Present). In C. Czosseck & K. Geers (Eds.), The Virtual Battlefield: Perspectives on Cyber warfare (pp. 182 - 188). IOS Press.
Geers, K. (2010). Live Fire Exercise: Preparing for Cyber War. Journal of
Homeland Security and Emergency Management, 7(1).
Hyacinthe, B. (2009). Cyber Warriors at War. Xlibris, pp. 82-85.
Hyacinthe, B. & Fleurantin, L. (2007). Initial supports to regulate information
warfare's potentially lethal information technologies and techniques.
Proceedings of the 3rd International Conference on Information War-
fare and Security (pp. 206-207). Academic Conferences Limited.
Kash, W. (2008). Lessons from the cyberattacks on Estonia. Retrieved from
http://gcn.com/articles/2008/06/13/lauri-almann--lessons-from-the-
cyberattacks-on-estonia.aspx.
Kaska, K., Talihärm, A.-M., & Tikk, E. (2010). Building a Comprehensive
Approach to Cyber Security. CCD COE Publications.
Landler, M., & Markoff, J. (2007). In Estonia, what may be the first war in
cyberspace. The New York Times. Retrieved from
http://www.nytimes.com/2007/05/28/business/worldbusiness/28iht-
cyberwar.4.5901141.html.
Lemay, A., Fernandeza, J. M., & Knight, S. (2010). Pinprick attacks, a lesser
included case? In C. Czosseck & K. Podins (Eds.), Conference on Cyber Conflict Proceedings (pp. 183 - 194). Tallinn: CCD COE Publications.
Liles, S. (2010). Cyber Warfare: As a form of low-intensity conflict and insur-
gency. In C. Czosseck & K. Podins (Eds.), Conference on Cyber Conflict Proceedings (pp. 47 - 57). Tallinn: CCD COE Publications.
MoD. (2004). National Security Concept of the Republic of Estonia.

MoD. (2008). Cyber Security Strategy. Retrieved from
http://www.mod.gov.ee/files/kmin/img/files/Kuberjulgeoleku_strateegia_
2008-2013_ENG.pdf.

MoD. (2010). NATIONAL SECURITY CONCEPT. Retrieved from
http://www.kmin.ee/files/kmin/nodes/9470_National_Security_Concept_
of_Estonia.pdf.

MoEAC. (2006). Estonian Information Society Strategy 2013. Retrieved from
http://www.riso.ee/en/system/files/Estonian Information Society Strategy 2013.pdf.

MoEAC. (2007). Implementation Plan 2007-2008 of the Estonian Information
Society Strategy.

MoEAC. (2009). Statute for the Development of National Information System
(in Estonian). Retrieved from https://www.riigiteataja.ee/akt/13219897.

MoFA. (2010a). Estonia Supports Council of Europe in Fight Against Cyber
Crime. Retrieved from http://www.vm.ee/?q=en/node/9315.

MoFA. (2010b). Foreign Minister Paet Invited EU and Southeast Asian Nations to Co-operate in Backing Cyber Defence. Retrieved from
http://www.vm.ee/?q=en/node/9512.

MoFA. (2010c). National Experts Shared Cyber Security Recommendations
with UN Secretary General. Retrieved from
http://www.vm.ee/?q=en/node/9722.

MoI. (2009). Estonian Emergency Act (unofficial translation). Retrieved
January 4, 2011, from
http://www.legaltext.ee/et/andmebaas/tekst.asp?loc=text&dok=XXXXX
26&keel=en&pg=1&ptyyp=RT&tyyp=X&query=h�daolukorra.

MoI. (n.d.). Ministry of the Interior, Department of crisis management and
rescue policy (in Estonian). Retrieved January 4, 2011, from
http://www.siseministeerium.ee/elutahtsad-valdkonnad-ja-teenused-2.

MoJ. (2010a). Guidelines for Development of Criminal Policy until 2018. Retrieved from http://www.just.ee/arengusuunad2018.

MoJ. (2010b). Estonian Code of Criminal Procedure (unofficial translation).
Retrieved from http://www.legaltext.ee/text/en/X60027K6.htm.

NATO. (2010). Strategic Concept for the Defence and Security of the Members of the NATO. Retrieved December 30, 2010, from
http://www.nato.int/cps/en/natolive/official_texts_68580.htm.

Nazario, J. (2007). Estonian DDoS Attacks – A summary to date. Retrieved
from http://asert.arbornetworks.com/2007/05/estonian-ddos-attacks-a-
summary-to-date/.

Nazario, J. (2009). Politically Motivated Denial of Service Attacks. In C.
Czosseck & K. Geers (Eds.), The Virtual Battlefield: Perspectives on
Cyber Warfare (pp. 163-181). 163-181: IOS Press.

Odrats, I. (Ed.). (2007). Information Technology in the Public Administration
of Estonia Yearbook 2007. Ministry of Economic Affairs and Communication.

Ottis, R. (2008). Analysis of the 2007 Cyber Attacks Against Estonia from the Information Warfare Perspective. Proceedings of the 7th European Conference on Information Warfare (p. 163). Academic Conferences Limited.

Ottis, R. (2009). Theoretical Model for Creating a Nation-State Level Offensive Cyber Capability. 8th European Conference on Information Warfare and Security (pp. 177-182). Academic Publishing Limited.

Tikk, E., Kaska, K., & Vihul, L. (2010). International Cyber Incidents: Legal Considerations (p. 130). Tallinn: CCD COE Publications.

Australian National Critical Infrastructure Protection: A Case Study

Matthew Warren and Shona Leitch
Deakin University, Australia

Originally Published in the Conference Proceedings of ECIW 2011

Editorial Commentary

Australia is a modern society and is highly dependent on key critical systems at the national and state level. Australia has developed sophisticated national security policies and physical security agencies to protect against current and future security threats associated with critical infrastructure protection and cyber warfare protection. This paper will review the current strategies used by Australia over a decade and evaluate their differences and discuss the reasons for these differences. Future threats such as Cyber Warfare and the steps that are being proposed will be considered. This paper will highlight current Australian best practices in critical infrastructure and cyber warfare protection many of which may be applicable in a European context and provide an informative contrast.

This paper will discuss some of the common security risks that face Australia and how their government policies and strategies have been developed and changed over time, for example, the proposed Australian Homeland Security department. This paper will discuss the different steps that Australia has undertaken in relation to developing national policies to deal with critical infrastructure protection.

Matthew Warren and Shona Leitch

Abstract: Australia has developed sophisticated national security policies and physical security agencies to protect against current and future security threats associated with critical infrastructure protection and cyber warfare protection. This paper will discuss some of the common security risks that face Australia and how their government policies and strategies have been developed and changed over time, for example, the proposed Australian Homeland Security department. This paper will discuss the different steps that Australia has undertaken in relation to developing national policies to deal with critical infrastructure protection.

Keywords: critical infrastructure, Australia and policy

1. Introduction

Australia is a modern society and is highly dependent on key critical systems at the national and state level. These key systems have become more dominant as the Information Age has developed. These key systems are grouped together and described as critical infrastructure; this is infrastructure so vital that its incapacity or destruction would have a debilitating impact on defence and national security (Lewis, 2006). Many of these critical systems are based upon ICT (Information and Communication Technology) systems.

Australia takes ICT security very seriously, it has been estimated that Australian organisations spend between A$1.37 – A$1.74 billion per year on IT security, and the total financial losses due to computer-related security incidents in the 2006 financial year have been estimated to be between $595 and $649 million (Australian Institute of Criminology, 2009).

This paper will review the current strategies used by Australia over a decade and evaluate their differences and discuss the reasons for these differences. Future threats such as Cyber Warfare and the steps that are being proposed will be considered. This paper will highlight current Australian best practices in critical infrastructure and cyber warfare protection many of which may be applicable in a European context and provide an informative contrast.

2. The initial view of the Australian Federal Government

The initial focus of the Australian Federal Government policy was that critical infrastructure protection was a commercial consideration and related to Information Security (Busuttil and Warren, 2004).The Australian Federal Government has been aware of the problems that Australian corporations may have with dealing with these new security issues. The Australian Federal Government has responded by offering advice for corporations. The initial Australian Government advice (AGD, 1998) suggested ways in which organisations could reduce Critical Infrastructure Protection risks (Busuttil and Warren, 2004):

- Organisations should implement protective security such as passwords etc in accordance to a defined security standard such as AS/NZS 4444 (Now 17799) (Information Security Management);
- Organisations should formally accredit themselves against security standards such as AS/NZS 4444 (17799);
- Organisations should raise awareness of security issues such as password security, E-commerce risks among their staff;
- Organisations should train their staff in how to use computer security systems efficiently and effectively.

This advice was subsequently updated and in 2004 the Australian Government responded with new security advice (Australian Government, 2004):

- The Australian and New Zealand Standard for Risk Management AS/NZS 4360:1999 is the standard by which all critical infrastructure will be assessed to assist with the review of risk management plans for prevention (including security), preparedness, response and recovery (PPRR).

In 2004 the Australian Federal Government formally defined the following; "Critical infrastructure is defined as those physical facilities, supply chains, information technologies and communication networks which, if destroyed, degraded or rendered unavailable for an extended period, would significantly impact on the social or economic well-being of the nation, or affect Australia's ability to conduct national defence and ensure national security" (Australian Government, 2004). In essence this description describes organisations that exist at a government level or at a corporate level (Australian Government, 2004).

Historically, much of Australia's infrastructure was originally owned and operated by the public sector at the federal, state and local government levels (Smith, 2004) however the majority of Australia's critical infrastructure has now been privatised and is under private sector ownership. Consequently, protecting Australia's critical infrastructure now requires a higher level of cooperation between all levels of government and the private sector owners. Hence, the federal government has developed a policy for critical infrastructure protection that focuses broadly on addressing the following strategies (Australian Government, 2004; AGD, 2004):

- Distinguishing critical infrastructures and ascertaining the risk areas;
- Aligning the strategies for reducing potential risk to critical infrastructure;
- Encouraging and developing effective partnerships with state and territory governments and the private sector;
- Advancing both domestic and international best practice for critical infrastructure protection.

As Warren and Leitch discussed (Warren and Leitch, 2010), the Australian Federal Government recognised the importance of crucial systems and the development of new industry support mechanisms, in particular Trusted Information Sharing Network (TISN).

The TISN is a forum in which the owners and operators of critical infrastructure work together by sharing information on security issues which affect critical infrastructure (TISN, 2007). TISN requires the active participation of Critical infrastructure Protection owners and operators of Critical infrastructure Protection, regulators, professional bodies and industry associations, in cooperation with all levels of government, and the public. To ensure this cooperation and coordination, all of these participants should commit to the following set of common fundamental principles of Critical infrastructure Protection (TISN, 2007). These principles are (TISN, 2007, Warren and Leitch, 2010):

- Critical infrastructure Protection is centred on the need to minimise risks to public health, safety and confidence, ensure economic security, maintain Australia's international competitiveness and ensure the continuity of government and its services;
- The objectives of Critical infrastructure Protection are to identify critical infrastructure, analyse vulnerability and interdependence, and protect from, and prepare for, all hazards;
- As not all critical infrastructure can be protected from all threats, appropriate risk management techniques should be used to determine relative severity and duration, the level of protective security, set priorities for the allocation of resources and the application of the best mitigation strategies for business continuity;
- The responsibility for managing risk within physical facilities, supply chains, information technologies and communication networks primarily rests with the owners and operators;
- Critical infrastructure Protection needs to be undertaken from an 'all hazards approach' with full consideration of interdependencies between businesses, sectors, jurisdictions and government agencies;

- Critical infrastructure Protection requires a consistent, co-operative partnership between the owners and operators of critical infrastructure and governments;
- The sharing of information relating to threats and vulner-abilities will assist governments, and owners and operators of critical infrastructure to better manage risk;
- Care should be taken when referring to national security threats to critical infrastructure, including terrorism, so as to avoid undue concern in the Australian domestic community, as well as potential tourists and investors overseas;
- Stronger research and analysis capabilities can ensure that risk mitigation strategies are tailored to Australia's unique critical infrastructure circumstances.

3. Australia's critical infrastructure – the alternative view point

During the time that the Australian Federal Government defined National Policy for Critical Infrastructure Protection, the opposition Australian Labor Party defined their own very different policy and viewpoint. The following is a time sequence of their policy development:

2001- Initial Policies

In October 2001, as a response to the act of terrorism in New York in September of the same year, the Australian Labor Party (ALP) which was the Australian opposition of the time led by Kim Beazley proposed a range of national security reforms. The reforms focussed on three main areas (ALP, 2001):

- Improving border security;
- Combating terrorism;
- Improving national security planning.

It was stated that these reforms were previously in the planning stages but the first announcement was made only two days after the attack on the World Trade Center. Australia's border security was announced by them as a high priority with changes to the coast guard and aviation security regimes upmost. In terms of the aviation industry this included more counter-terrorism measures including the Federal Government taking over responsibility for all airport security checks, making sure there is a visible presence of officers at airports and introducing tighter controls on aviation security information by amending current laws and regulations (ALP, 2001).

Rather than just concentrating on the physical security controls which were very much in the forefront of the public's mind in 2001, the ALP also proposed a range of changes and initiatives in regards to protecting Australia's national infrastructure. This focused in on the establishment of a Defence Cyber-warfare Task Force which would use all the elements and agencies of the current defence force to counteract cyber security threats and cyber terrorism attacks.

This strategy proposed by the ALP revolved around the concept of "Homeland Security", this was the first time this term was used in Australia and the notion of integrating security agencies, expanding the range of activities and including the national infrastructure (transport, electricity, water, communication systems etc) as the most important elements was a dramatic leap in the protection of Australia from cyber terrorist threats.

2003 – 2005 – A Period of Reflection

In 2003, the Australian Labor Party was still the countries opposition party. The department of Homeland Security was still being advocated by them, so much so that an opposition Security minister was created whose portfolio encompassed border protection, crime prevention, intelligence-gathering, investigation and prosecution

(ALP, 2003) and they set up the Shadow Department of Homeland Security Portfolio.

In 2005 the notion of a Homeland Security department was still forefront in the ALP's policies as a way to address the issues of national security and bring together all of Australia's defence agencies (as was done during the 2000 Olympic Games held in Sydney, Australia). They believed that this level of integration and cohesion was the only way to truly protect Australia and its citizens from the continued threats and attacks. They outlined a number of cases which they felt supported this proposal (Beazley, 2005):

- The alleged involvement of Sydney Airport baggage handlers in an international drug trafficking syndicate. The Australian Federal Police claims baggage handlers were key players in a conspiracy to smuggle cocaine worth $15 million into Australia;
- Constant warnings from the Transport Workers' Union, that the Federal government had been aware of potential security breaches at Australian airports for at least four years and the TWU's call for improved security checks of short term employees and the immediate x-ray screening of all baggage and freight;
- Passengers' baggage containing large amounts of narcotics being diverted to domestic carousels to avoid Customs inspections;
- 39 security screeners out of 500 employed at the airport have serious criminal convictions, with a further 39 convicted of minor matters;
- Engineers with unauthorised duplicate keys;
- Lack of customs checks on airline staff.

2007 – 2008 From Opposition to Government

The Australian Labor Party in 2007 had moved from being the opposition party to forming Government. One of their main policies leading into the Federal election was that they would continue with their long term plan of forming a Department of Homeland Security.

In 2008, the Prime Minister announced that he planned to cancel the long term plans of the ALP to create the new department on the basis that the integration of all the defence agencies would be too "cumbersome" (Franklin and Walters, 2008).

Seven years of planning and proposals had disappeared less than a year after an election due to the complexities of how administration would be dealt with and confusion over how the complex integration could be achieved (Nicholson, 2008).

The fact that the initial plans arose swiftly after the terrorist attacks in September 2001 may pose questions as to whether the plans were ill thought out and borne out of the need to react rather than a sensible, productive and workable policy.

4. Recent Australian Government strategy

The Australian Federal Government (2008) has identified new security challenges, `it is increasingly evident that the sophistication of our modern community is a source of vulnerability in itself. For example, we are highly dependent on computer and information technology to drive critical industries such as aviation; electricity and water supply; banking and finance; and telecommunications networks. This dependency on information technology makes us potentially vulnerable to cyber attacks that may disrupt the information that increasingly lubricates our economy and system of government` (Rudd, 2008). This public acknowledgement by the Australian Prime Minister, Kevin Rudd, identifies the new security challenges facing critical infrastructure protection and highlighted the following security concerns (Rudd, 2008):

- Maintaining Australia's territorial and border integrity;
- Promoting Australia's political sovereignty;

- Preserving a cohesive and resilient society and strong economy;
- Protecting Australians and Australian interests both at home and abroad, and
- Promoting a stable, peaceful and prosperous international environment; particularly in the Asia-Pacific region, together with a global rules-based order which enhances Australia's national interests.

In 2009 the Federal Australian Government has responded to the issues regarding cyber security and critical infrastructure by proposing a coherent and government led approach to critical infrastructure protection. The primary objectives identified focus on all areas of Australian society where there are security risks, e.g. that individuals should be aware and take steps to "protect their identities, privacy and finances online" (Australian Government, 2009) that businesses and the government operate "secure and resilient information and communication technologies" and trusted electronic operating environment that supports Australia's national security and maximises the benefits of the digital economy (Australian Government, 2009). The Australian Federal Government also has developed a wide range of new strategic directions to focus Australia's cyber security programs (Australian Government, 2009):

- Improve the detection, analysis, mitigation and response to sophisticated cyber threats, with a focus on government, critical infrastructure and other systems of national interest;
- Educate and empower all Australians with the information, confidence and practical tools to protect themselves online;
- Partner with business to promote security and resilience in infrastructures, networks, products and services;
- Model best practice in the protection of government ICT systems, including the systems of those transacting with government online;

- Promote a secure, resilient and trusted global electronic operating environment that supports Australia's national interest;
- Maintain an effective legal framework and enforcement capabilities to target and prosecute cyber crime;
- Promote the development of a skilled cyber security workforce with access to research and development to develop innovative solutions.

As part of the new Australian Federal Government strategy, a new of number bodies have been developed with new capabilities. These include (Australian Government, 2009):

- CERT (Computer Emergency Response Team) Australia;
- This new Government body has moved to a national level to enable a "more integrated, holistic approach to cyber security across the Australian community";
- Some of the previously formed cyber security activities that were undertaken by numerous different agencies such as the Australian Government's Computer Emergency Readiness Team (GovCERT) have been combined together to form CERT in order to promote a greater (shared) understanding; provide targeted advice and give Australians a single point of contact.
- Cyber Security Operations Centre (CSOC).
- The core functions of the CSOC are focused mainly on government, infrastructure and critical private sector systems and aims to be a source for all issues related to awareness (especially the detection of sophisticated threats) and a facility to respond to cyber security risks and problems which are of national importance.

Another key aspect of CSOC is that it provides Australian Defences with a cyber warfare capability and provides a resource designed to service all government agencies (DSD, 2011).

The Australian Federal Government has started to refocus away from Critical Infrastructure Protection to Critical Infrastructure Resilience. The Australian Attorney General Robert McClelland announced that "The time has come for the protection mindset to be broadened – to embrace the broader concept of resilience". The aim is to build a more resilient nation – one where all Australians are better able to adapt to change, where we have reduced exposure to risks, and where we are all better able to bounce back from disaster" (TISN, 2010).

The Australian Federal Government in 2010 launched the *new Critical Infrastructure Resilience Strategy*. The aim of this new strategy is the continued operation of critical infrastructure in the face of all hazards as this critical infrastructure supports Australia's national defence and national security and underpins our economic prosperity and social wellbeing. More resilient critical infrastructure will also help to achieve the continued provision of essential services to the community (Australian Government, 2010). This new strategy also deals with new areas such as disaster protection and disaster resilience and this shift in policy is going to have a major impact upon Australia.

5. Discussion

The major issue facing Australia is the currently adopted distributed model of critical infrastructure protection and decision making and how that can effectively manage and secure Australia's critical infrastructure. Whilst it is commendable that an Australian Federal Government has faced the issue of critical infrastructure and cyber threats, the fact that this approach attempts to cover the entirety of Australia may in itself be problematic. There has been some streamlining of operations by nationalising CERT, however there are still a number of separate agencies that are involved in this process; Attorney-General's Department (AGD), the Australian Communications and Media Authority (ACMA), the Australian Federal Police (AFP), the Australian Security Intelligence Organisation's (ASIO), the

Defence Signals Directorate (DSD), the Department of Broadband, Communication and the Digital Economy (DBCDE), the Australian Government Information Management Office (AGIMO), the Joint Operating Arrangements (JOA) and the Cyber Security Policy and Coordination (CSPC) Committee. It is clear that there has been an overall Government shift to form two "umbrella" agencies (CERT and CSOC) to monitor, promote and control cyber threats however complexity will still arise as there as so many sub agencies that are involved in this process. In an area such as cyber security where speed is often of upmost importance to limit damage, the interaction of a large number of other agencies will surely slow this process down. If an Cyber attack occurs in real time against Australia, would they be able to react and make decisions in real time, or would the distributed model actually impact the decision making process? Another unique issue that relates to Australia is the federated government system consisting of a federal government and a number of state governments. A key issue is that when an attack occurs against an infrastructure at a state level that the response time to escalate the decision making process to the Federal government may be slow. This time lag could cause serious consequences and limit the effectiveness of these agencies.

A new factor with the introduction of CSOC is the move away from civilian organisations protecting Australia's critical infrastructure and cyber security risks and making defence organisations responsible for this role. This may heighten the chance of attacks against Australia's critical infrastructure because it could be considered a military target. The major shift in Australian policy is the announcement in 2010 of the move away from Critical Infrastructure Protection to Critical Infrastructure Resilience and the inclusion of natural disaster into the policy. This will have a major impact upon Australia and the real implications are still yet to emerge especially with the recent natural disasters in Australia.

6. Conclusion

Australia over the last decade has taken major steps in the protection of its national critical infrastructure. The Australian model is a workable model that has helped to protect Australian critical infrastructure against physical and cyber risks. The issue is whether the distributed model will work in a real time situation and whether the time delays would impacts the decision making processes.

A new emerging issue is the focus upon Critical Infrastructure Resilience and the future impact that this may have.

References

Australian Labor Party (ALP) (2001). Labor's Better Plan For Defence - A Secure Future, Canberra.

Australian Labor Party (ALP) (2003). ALP News Statement – Development of Homeland Security Portfolio, Canberra.

Attorney-General's Department (AGDs) (1998). Report of the Interdepartmental Committee on Protection of the National Information Infrastructure, Available from: http:// law.gov.au/publications/niireport/niirpt.pdf, visited 10th March, 2007.

Attorney General Department (AGD) (2004). Critical Infrastructure Protection National Strategy, Available from: http://www.nationalsecurity.gov.au, Accessed 10th November, 2007.

Australian Government (2004). Protecting Australia Against Terrorism, Department of the Prime Minister and Cabinet, Barton, ACT.

Australian Government (2009). Cyber Security Strategy, Attorney Generals Department, Commonwealth of Australia, ISBN 978-1-921241-99-4.

Australian Government (2010). Critical Infrastructure Resilience Strategy, Attorney Generals Department, Commonwealth of Australia, ISBN: 978-1-921725-25-8.

Australian Institute of Criminology (2009). Australian Business Assessment of Computer User Security, ISBN 978 1 921532 35 1.

Beazley, K. (2005). A Nation Unprepared: Australia in the Fourth Year of a Long War, Address to the Sydney Institute, Sydney, 4th August.

Busuttil, T. and Warren, M. (2004). A risk analysis approach to critical information infrastructure protection, Proceedings of the 5th Australian Information Warfare and Security Conference, Perth, Western Australia.

Defence Signals Directorate (DSD) (2011). CSOC - Cyber Security Operations Centre, Available from: http://www.dsd.gov.au/infosec/csoc.htm, Accessed 10th January, 2011.

Franklin, M and Walters, F (2008). Homeland Security Division Faces Axe, The Australian, May 8th.

Lewis, T (2006). Critical Infrastructure Protection in Homeland Security, Wiley Publishers, USA, ISBN 978-0-471-78628-3.

Nicholson, B. (2008). PM abandons "cumbersome" homeland security department. The Australian, December 4th.

Smith, S. (2004). Infrastructure, [Online], NSW Parliament, Available from http://www.parliament.nsw.gov.au/prod/parlment/publications.nsf/0/C63 89C30B0383F9ACA256ECF0006F610, Accessed 10th November, 2009.

TISN (Trusted Information Sharing Network) (2007). About Critical Infrastructure, Available from: http://www.tisn.gov.au, Accessed, 15th July, 2009.

TISN (Trusted Information Sharing Network) (2010). The Shift To Resilience, CIR News, Vol 7 &, No 1.

Rudd, K. (2008). The First National Security Statement to the Parliament Address by the Prime Minister of Australia, The Hon. Kevin Rudd MP, URL: http://www.pm.gov.au/media/speech/2008/speech_0659.cfm, Accessed, 10th December, 2008.

Warren, M. and Leitch. S. (2010). Commercial Critical Systems and Critical Infrastructure Protection: A Future Research Agenda, Proceedings of the 2010 European Information Warfare Conference, Thessaloniki, Greece.

Proactive Defense Tactics Against On-Line Cyber Militia

Rain Ottis
Cooperative Cyber Defence Centre of Excellence, Tallinn, Estonia
Originally Published in the Conference Proceedings of ECIW 2010

Editorial Commentary
The paper reflects on the developing trend of "popular" cyber campaigns that mirror political, economic or military cyber conflicts in cyberspace. The Estonian cyber arrack case from 2007 showed that a whole nation-state can be affected by cyber attacks, whereas the Georgian cyber attacks of 2008 is an illustration of a cyber campaign that mirrors an armed conflict. In both cases at least part of the attacks were likely committed by patriotic hackers – volunteers who use cyber attacks to take part in intra- or international conflicts. In such cyber conflicts usually only the targets are known while the aggressors remain anonymous. The paper provide a theoretical overview of a specific type of on-line cyber militia and then propose tactics to neutralize them and their activities. The tactics are based on a proactive defense posture and primarily use information operation techniques to achieve the effect from within the cyber militia itself and are discussed within the paper.

Abstract: There is a developing trend of "popular" cyber campaigns that mirror political, economic or military conflicts in cyberspace. The Estonian case from 2007 showed that a whole nation-state can be affected by cyber attacks, whereas the Georgian case of 2008 is an illustration of a cyber campaign that mirrors an armed conflict. In both cases at least part of the attacks were likely committed by patriotic hackers – volunteers who use cyber attacks to take part in intra- or international conflicts. In such cyber conflicts usually only the targets are known while the aggressors remain anonymous. It is often difficult to discern where state capability ends and independent patriotic hacker groups begin. Furthermore, it is relatively easy to form a new cyber militia from people who have little prior experience with computers. I define *cyber militia* as a group of volunteers who are willing and able to use cyber attacks in order to achieve a political goal. I further define *on-line cyber militia* as a cyber militia where the members communicate primarily via Internet and, as a rule, hide their identity. What the newly-minted cyber warriors may lack in skill and resources, they can often compensate with numbers. However, even an ad-hoc cyber militia that is not under direct state control can be a useful extension of a state's cyber power. On the other hand, they can also become a threat to national security. Due to the global nature of the Internet, this threat is most likely coming from multiple jurisdictions, which limits the law enforcement or military options of the state. Therefore, other approaches should be considered. In order to understand the potential threat from cyber militias, either ad-hoc or permanent, we need to explore how they are organized. I provide a theoretical overview of a specific type of on-line cyber militia and then propose tactics to neutralize it. The tactics are based on a proactive defense posture and primarily use information operation techniques to achieve the effect from within the cyber militia itself.

Keywords: cyber conflict, cyber militia, proactive cyber defense, information operations, hactivism

1. Introduction

Over the past few decades the malicious activity in cyberspace has grown to levels, where it is now considered a national security issue. This is arguably due to the fact that computers have become nearly ubiquitous in modern societies. They are easy to use and very accessible, allowing the people to regularly communicate, learn, work and have fun in cyberspace (Ottis 2010).

On the other hand, it is now also easier to use this technology for malicious purposes. There are automated cyber attack kits, vulnerability databases and instruction manuals for conducting offensive operations in cyberspace. The skill level of the attacker today does not need to be high. On the contrary, some of the more visible attacks are often perpetrated by individuals with little or no computer training (Carr 2009).

While cyber crime continues to thrive in the quest for illegitimate income via cyberspace operations, the politically motivated attacks are becoming ever more common and visible. Many international conflicts in recent years have had a mirror campaign in cyberspace. The question that often develops is whether or not the cyber campaign is sponsored by the state(s) involved in the conflict, as the attacks usually seem to be the work of patriotic hackers. (Carr 2009, Nazario 2009)

However, it is quite possible that even without a direct command link with the state, the attackers still act according to the state's agenda. After all, the state may use this volunteer force in order to maintain deniability. The official cyber warriors (military, intelligence etc.) of the state are just one of the potential components of a national offensive cyber capability. Volunteers (patriotic hackers, hactivists) and mercenaries (criminals, commercially hired experts etc.) can augment the organic cyber capabilities of the government. (Ottis 2009)

People can also mobilize as a result of a true grass roots movement. Such independent groups could organize a cyber attack campaign as a sign of protest or to promote their views. If compared to an entity that has hidden state sponsorship, they would most likely look very similar to an outside observer. Either way, this type of on-line group can evolve into a threat beyond mere inconvenience as seen in cases like the Estonian and Georgian cyber conflicts. (Ottis 2008, Carr 2009, Nazario 2009, Denning 2010)

In order to cope with this threat, we must first understand how it works. Therefore, I will provide a theoretical overview of the organizational aspects of non-state political activist groups who use cyber attacks and then look at some tactics to counter these groups.

2. On-line cyber militia

Denning (2010) describes three categories of non-state attackers (separate from the ordinary cyber criminal): patriotic hackers, electronic jihadists and hactivists. The main difference among them is the choice of targets, although Denning admits that they could all be lumped together as hactivists. For the purposes of this research, however, this distinction does not matter, as the focus is on how they are organized, not who they are fighting for or against. In particular, I am interested in finding potential weaknesses in the organization and operation of cyber militias.

Let us define *cyber militia* as a group of volunteers who are willing and able to use cyber attacks in order to achieve a political goal. Let us further define *on-line cyber militia* as a cyber militia where the members communicate primarily via Internet and, as a rule, hide their identity (for example, by using a hacker alias). Cyber militias can be ad-hoc (gathering only for a specific occasion) or permanent.

The word "volunteers" in the definition refers to people who participate in the cyber militia of their own free will. They do not get paid for their activities, nor do they have a contractual obligation to the militia. They have the right to choose their level of commitment and to leave the militia, if and when they wish. Therefore, volunteer soldiers who join a government run cyber attack unit are not considered a cyber militia.

The word "political" in the definition refers to all aims that transcend the personal interest of the volunteer. This includes religious views, nationalistic views, opinions on world social order etc.

In the context of this analysis, I am focusing on a subset of on-line cyber militias that meet the following criteria:

- The communication within the militia is centralized
- There is no direct state support or control of the militia

- The members are loosely connected in real life

The centralized communication constraint is a fairly standard arrangement for communicating, preparing, planning and coordinating a cyber attack campaign of the cyber militia. Perhaps the most used communication channels are on-line forums and instant messaging services. (Carr 2009, Denning 2010) This is also very useful for the defending side, especially for observing, infiltrating and neutralizing the cyber militia.

A cyber militia that receives direct support or instructions from the government should be considered as an organic component of the state and is therefore outside the scope of this research. However, indirect or covert state support or control (as long as it is not well known among the militia) remains still in the area of interest.

Although the leadership or core group in a militia probably is personally acquainted, as a whole the members of the on-line cyber militia are loosely connected in real life. In this case loosely connected means that most members know no or few other members and nobody knows the entire membership in person. This requires them to communicate over the Internet and coincidentally makes them more susceptible to information operations techniques. While this constraint is not true in every case, it should be a safe assumption in large (numbering in the hundreds) militias and can also hold in smaller organizations.

From the forum posts it should be possible to identify the roles of the people in the cyber militia. Key "officer" roles include leaders, trainers, suppliers, while the rest could be categorized as soldiers, and "camp followers". The leaders provide motivation for action, coordination of effort and direction of attacks. The trainers provide instructions for reconnaissance, attack and covering tracks. Suppliers provide tools, such as scanners, attack kits and malware. Soldiers participate actively in the attacks, but can be expected to remain relatively passive on the forum, potentially reporting attack results or targeting information. Camp followers read the forum for their own interest, but do not participate in the planning or execution of attacks. Identifying the different roles in the organization offers individual targeting opportunities as well as potential avenues for infiltration.

Since the cyber militia is not necessarily a formal organization, the same person may have several roles, which can change over time. It is also important to note that an "officer" role is often not appointed by the militia, but acquired by the member by actively participating in the activities.

3. Neutralizing an on-line cyber militia

Assuming that on-line cyber militias can be a considerable threat to national security, there should also be ways of neutralizing this threat. Using traditional law enforcement methods or military force is often not feasible, because personal attribution is seldom achieved and the militia members can reside in a number of different unfriendly and uncooperative jurisdictions. Therefore I will consider alternative tactics of neutralizing an on-line cyber militia. In particular, I will propose options from the strategic starting point of information operations and proactive defense.

An important caveat here is that I do not presume universal legality of any of the tactics. It would be very difficult to do, given that the legal status of the cyber militia and its actions may vary greatly, depending on the case. For example, the cyber militia may act completely within the legal framework of the host state. On the other hand, militia members could be considered illegal combatants who may be targeted for military action (Schmitt 2002). Therefore, the tactics below should be considered as theoretical options only, not as a policy manual for dealing with a cyber militia.

There are two points where the activity of an on-line cyber militia is potentially visible for observation. First, there are the logs at the targeted sites. Second, the shared communication channel (a forum, for example) where they gather, exchange opinions and plan their activities. The two places where the militia is visible are also the places where one can fight them.

Sun-Tzu said: "Thus the highest realization of warfare is to attack the enemy's plans; next is to attack their alliances; next to attack their army; and the lowest is to attack their fortified cities" (cited in Sawyer 1994). I will use this principle as a loose framework for considering tactics. The analogies do not need to be an exact fit and should be interpreted liberally. First, I will look at how to neutralize the militia's ability to plan and coordinate attacks. Second, I will look at ways of attacking the virtual alliances between the members that make up the cyber militia. Third, I will look at neutralizing the effectiveness of the militia's cyber attacks. Last, I consider a counterattack against the actual communication service that is the heart of the militia's operations.

It is important to note that for the countermeasures to work, it is necessary to gain access to the main communication channel of the militia. This may be as simple as monitoring a public forum, but a more likely scenario would require at least some form of infiltration into the channel. The infiltration does not need to be very deep - a "soldier" level access would likely be sufficient to gather the necessary information about the militia. Infiltration is required, because any sufficiently mature cyber militia will likely try to hide or protect itself from outside entities. For example, the StopGeorgia.ru forum blocked US-based IP addresses to stop researchers from accessing the forum during ten days in August of 2008 (Carr 2009).

3.1. Attacking plans

One way to neutralize the militia can be called *poisoning the well* tactic. It refers to corrupting the shared communication channel with de-motivational posts, self-destructive or ineffective attack tools and methods, bad targeting data, etc. As a result, the channel loses its effectiveness as a means for coordinating the actions of the militia, the members grow frustrated with apparent lack of coherence, and the aggression gets released inside the militia in the form of angry debate. If the militia is perceived as ineffective by the members, it will eventually disband.

An alternative approach would be to hijack the militia by shifting the debate to attacking other targets. This would basically deflect the blow from the original target, making it safe.

Yet another approach is to carry out an attack in the name of the militia against a powerful third entity in order to provoke a counter-strike against the entire militia (a false flag attack). In other words, pull a strong opponent into the fight, forcing the militia into defensive positions. As a result, the militia will have to drop its plans for the original target.

3.2. Attacking alliances

Presumably, members of the militia want to remain *anonymous* and would leave or become inactive if there was a serious chance of being personally identified. This presents another opportunity to disband the militia from within by breaking the virtual alliances between militia members.

Without attribution there can be no personal consequences. On the other hand, if the anonymity is lost (or perceived lost by the membership), the militia will lose its trustworthiness. As a result, the militia will either disband or search for an alternative (clean) communication channel. However, since the infiltrated agents will also move over to the new channel, it would only be a temporary solution.

The question is, then, how to identify the members of the forum. In reality, it is probably not necessary to identify all or even most of the members. Most likely it is enough to break the cover of one or a

few people, in order to create mistrust and fear of real life consequences in a considerable portion of the membership.

There are many ways to potentially achieve attribution of a few individuals. The simplest is to "break the cover" on infiltrated agents
(can use fake identities, as they would be difficult to verify by other
members) and have them "confirm" it. Another is to offer attack
tools to the forum that provide the information that is necessary for
personal attribution (basically a Trojan). Yet another is to correlate
target log data with forum posts, and go through the legal channels.
Of course, attribution may be achieved by simply arranging a meeting in real life.

Note that it may not be necessary to actually follow up the attribution with legal or military action. Just posting the personal details of
some users on the forum could be enough to make a considerable
portion of the members leave.

3.3. Attacking the army

The loose analogy to an army in this case could be the cyber attacks
organized by the militia (the soldiers that have marched to the city
gates). Obviously, the defensive actions at the target come from the
long list of standard cyber security measures. However, these can
be deployed much more effectively, if the infiltrated agent can relay
the attack plans to the defenders. Knowing when, where and how
the attack will come makes the work of defenders much easier and
blunts the effectiveness of the attackers. This, in turn, may have a
demoralizing effect on the militia.

3.4. Attacking fortified cities

If we take the forum to be the fortified city that serves as a home
base for the cyber militia, then obviously there are ways of attacking
it. Conceptually the easiest would be to use law enforcement to
have it taken down, or if that fails, launch a denial of service attack
against the server that hosts the service. Alternatively, one could

take over and shut down the forum with hacking techniques. The problem with this approach is that the militia can easily regroup using a secondary meeting point (for example, a pre-determined IRC channel or a website). In addition, the counterattack will likely motivate them to continue the fight, as it is now a more personal matter. Therefore, this option, while potentially the easiest to achieve, is also least likely to generate a lasting effect.

In addition, it would be possible to post messages and materials in the channel that are against the enforced laws in the jurisdiction (vs posting attack instructions, which may be illegal but not enforced by a militia-friendly government), thus provoking a collateral response from the Internet service provider or law enforcement community.

4. Limitations and future research

All views in this work are attributed to the author and should not to be considered as the views or policy of the Cooperative Cyber Defence Centre of Excellence or the North Atlantic Treaty Organization.

The analysis of the on-line cyber militia is based on existing literature and provides only a theoretical viewpoint to the problem area. It is a generalized model, which may not apply to all cases.

The outlined tactics are a blend of information operation techniques and offer a more proactive defense posture against an on-line cyber militia. However, depending on the case, they may contain elements that can be considered against the law. Issues that may arise include perfidy, freedom of speech, computer crime, illegal surveillance and personal data protection, to name a few. Therefore, it should not be viewed as a policy manual, but a theoretical overview of alternative tactics.

One way to move this research forward is to use social network analysis on cyber militia forum logs to identify the roles and interactions of key actors in cyber militia. This could help develop a better model for a generic cyber militia and a more detailed method for targeting key members in the militia.

5. Conclusion

The rising trend of politically motivated cyber attacks by non-state actors has changed the balance of power in cyberspace. On-line cyber militia is a type of organization that allows everyone with a computer and an Internet connection to become active in the world of cyber conflict.

Since on-line cyber militias have shown the capability to become a threat to national security, it is important to study them. I have given an theoretical overview of a specific type of cyber militia, which relies on a mass of anonymous members for its "firepower". More importantly, I have provided some generic tactics for neutralizing the on-line cyber militia, under the strategic approach of information operations and proactive defense.

References

Carr, J. (2009) Inside Cyber Warfare, Sebastopol, CA: O'Reilly Media.
Denning, D. E. (2010) "Cyber Conflict as an Emergent Social Phenomenon", Corporate Hacking and Technology-Driven Crime: Social Dynamics and Implications (Hold, T. & Schell, B. eds.), IGI Global. [to appear]
Nazario, J. (2009) "Politically Motivated Denial of Service Attacks", The Virtual Battlefield: Perspectives on Cyber Warfare (Czosseck, C. & Geers, K. eds.), Amsterdam: IOS Press, pp 163-181.
Ottis, R. (2008) "Analysis of the 2007 Cyber Attacks Against Estonia from the Information Warfare Perspective", Proceedings of the 7th European Conference on Information Warfare and Security, Plymouth, Reading: Academic Publishing Limited, pp 163-168.
Ottis, R. (2009) "Theoretical Model for Creating a Nation-State Level Offensive Cyber Capability." Proceedings of the 8th European Conference on Information Warfare and Security, Lisbon, Reading: Academic Publishing Limited, pp 177-182.
Ottis, R. and Lorents, P. (2010) "Cyberspace: Definition and Implications." Proceedings of the 5th International Conference on Information Warfare and Security, Dayton, US. [accepted for publication]

Schmitt, M. (2002) "Wired Warfare: Computer Network Attack and International Law", International Review of the Red Cross, Vol 84, No 846, pp 365-399.
Sawyer, R.D. (1994) Sun-Tzu: The Art of War, Boulder: Westview Press

Analysis of the 2007 Cyber Attacks Against Estonia from the Information Warfare Perspective

Rain Ottis
Cooperative Cyber Defence Centre of Excellence, Tallinn, Estonia
Originally Published in the Conference Proceedings of ECIW 2008

Editorial Commentary

Following the relocation of a Soviet-era statue in Tallinn in April of 2007, Estonia fell under a politically motivated cyber attack campaign lasting twenty-two days. Perhaps the best known attacks were distributed denial of service attacks, resulting in temporary degradation or loss of service on many commercial and government servers. While most of the attacks targeted non-critical services like public websites and e-mail, others concentrated on more vital targets, such as online banking and DNS. At the time of this writing – more than six months after the cyber attacks – no organization or group has claimed responsibility for the cyber attacks, although some individuals have been linked with carrying them out.

This paper will argue that the key to understanding the cyber attacks that took place against Estonia in 2007 lies with the analysis of an abundance of circumstantial evidence that ran parallel to the cyber attacks. These consisted of political,

economic and information attacks on Estonia, as well as isolated cases of physical violence. Clear political signatures
were even detected in the malicious network traffic. All told,
it is clear that the cyber attacks were linked with the overall
political conflict between Estonia and Russia.

Abstract: Following the relocation of a Soviet-era statue in Tallinn in April of 2007, Estonia fell under a politically motivated cyber attack campaign lasting twenty-two days. Perhaps the best known attacks were distributed denial of service attacks, resulting in temporary degradation or loss of service on many commercial and government servers. While most of the attacks targeted non-critical services like public websites and e-mail, others concentrated on more vital targets, such as online banking and DNS. At the time of this writing – more than six months after the cyber attacks – no organization or group has claimed responsibility for the cyber attacks, although some individuals have been linked with carrying them out.

This paper will argue that the key to understanding the cyber attacks that took place against Estonia in 2007 lies with the analysis of an abundance of circumstantial evidence that ran parallel to the cyber attacks. These consisted of political, economic and information attacks on Estonia, as well as isolated cases of physical violence. Clear political signatures were even detected in the malicious network traffic. All told, it is clear that the cyber attacks were linked with the overall political conflict between Estonia and Russia.

While some analysts have considered last year's events in Estonia an international, grass roots, display of public opinion, there are some direct and many indirect indications of state support behind what can be best described as an information operation. By information operation, the author means the use of information and information technology to affect the decisions and actions of an opponent.

The paper will give an overview of the major events and provide an analysis of the attacks from the information warfare perspective. The paper will also discuss some of the potential problems with using the Internet as a field of battle by lone hackers, terrorist groups and states. To a minor degree, the paper will also cover the difficulties associated with investigating and analyzing international cyber attacks. The objective of this paper is not to implicate a specific organization or entity, but to provide a wider view to the cyber attacks that were carried out against Estonia in the spring of 2007.

Keywords: cyber attack, information operation, people's war

1. Introduction

In the spring of 2007 Estonia fell under a cyber attack campaign lasting a total of 22 days. The attacks were part of a wider political conflict between Estonia and Russia over the relocation of a Soviet-era monument in Tallinn. Due to the lack of definitive quantitative data, the author will use qualitative analysis to explain the cyber attacks.

1.1. The trigger

The trigger for the event was the Estonian government's decision to relocate a monument to Soviet troops from a busy intersection in central Tallinn to a nearby military cemetery. The monument depicting a Soviet soldier was originally erected in 1947 at the burial site of Soviet troops who died while taking Tallinn in World War II. Since that time the monument has developed two very distinct identities. For the local Russian minority it represents the "liberator" while for the Estonians it represents the "oppressor".

Over the past few years the statue had become a focal point of tension between pro-Kremlin and Estonian nationalist movements. In order to defuse the situation and to relocate the war-dead from a traffic intersection to a more peaceful resting place the Estonian government decided to move the monument and the accompanying remains to a military cemetery in Tallinn. Work began on the 26th of April 2007. During the day, mostly peaceful protesters gathered at the site, but in the evening a more violent crowd emerged. After a few hours of violent clashes with the police the rioters turned away and proceeded to vandalize and loot the nearby stores. Police regained control of the situation by morning.

However, the 27th of April marked the beginning of cyber attacks that targeted Estonian internet-facing information systems. Attacks of various types continued for a total of 22 days. Even though the attack types were well known, they were unparalleled in size and variety compared to a country the size of Estonia. Furthermore, Estonia is highly networked, so a wide scale attack on the availability of public digital services has a significant effect on the way of life of ordinary citizens and businesses alike. Therefore, these cyber attacks cannot be disregarded as mere annoyances but should be considered a threat to national security.

1.2. Overview of associated events

During the 27th of April there were several smaller standoffs between rioters and police. Throughout this time, both local and international media reported on the street riots. Interestingly, the looting of the stores and destruction of property was not covered by Russian media, who mostly reported on police violence against "peaceful protesters". This fueled an array of angry articles and statements from Russia, including a statement by a member of the Russian parliament that this event should be cause for war (ICDS 2007). It is therefore understandable why many Russians could be inclined to participate in various actions against Estonia.

Aside from the cyber attacks, the most notable events transpired at the Estonian embassy in Moscow. Pro-Kremlin youth groups staged well organized and equipped protests for many days and at times actually prevented Estonian embassy workers and diplomats from entering or exiting the building. The climax came on May 2nd, when the Estonian ambassador was physically attacked during a press conference. (ICDS 2007)

Another aspect of the conflict was economical. While officially no economic sanctions were imposed on Estonia by Russian authorities, the trade relations deteriorated. Many companies in Estonia lost revenue with Russian trade. This could be explained as a patriotic reaction by the business owners in Russia. On the other hand, the sudden ban on heavy commercial truck traffic at a border bridge in Narva clearly required Russian government involvement (Ottis 2007). The ban was lifted when the situation calmed down.

2. Facts

For this analysis, the author was able to re-use facts gathered for an earlier analysis of the same event (Ottis 2007). In addition, updated information from the Estonian State Procurature is included.

2.1. Facts collected by the author during and after the events in question

- The cyber attacks in question took place between 27 April and 18 May of 2007. The focus, method and volume of the attacks shifted during this period, but most of the detected attacks can be attributed to the same underlying event.
- The vast majority of the malicious traffic originated from outside Estonia. To combat this, some banks temporarily cut off all foreign traffic while remaining accessible for clients in Estonia. This white list was then gradually expanded to include the countries with many clients but few attackers.
- The malicious traffic often contained clear indications of political motivation and a clear indication of Russian language background. For example, malformed queries directed at a government website included phrases like *"ANSIP_PIDOR=FASCIST"* (Mr. Ansip was the Estonian Prime Minister at the time). Dozens of variants were used, often containing profanities.
- Instructions for attacking Estonian sites were disseminated in many Russian language forums and websites. These instructions often included motivation, targeting and timing information, as well as a specific description for launching attacks. An example of these instructions is displayed in Figure 1. Note that this excerpt includes information about when, what and how to attack. It also illustrates how simple the most primitive attacks are to organize, provided you can motivate enough people to execute these simple instructions. With thousands attacking, even a primitive ping flood can cause trouble.

На **9-е МАЯ** планируется повтор данной акции!
**Не дай унизить своих соотечественников, отомсти за
издевательства !!!**
@ адреса е**SS**тонских депутатов

Программа для рассылки писем
(пароль на RAR: nnm)

Нажми (**пуск -> выполнить -> cmd**)
введи ping -n 5000 -l 10000 э**SS**тонский_сайт -t . и жми **ENTER** ВСЕ !!! Твои пламенные
приветы полетели...
пример: ping -n 5000 -l 1000 www.riik.ee -t
Это 3 элементарных действия, после которых многие эстонские сайты просто перестанут
работать!!!
Или вот .BAT файл, который в автоматическом режиме последовательно пингует эстонские DNS и
MAIL сервера. Цикл бесконечен :)
Скопировать (красным) нижеприведённый текст, вставить в блокнот и сохранить как
priveteSStonia.BAT (название можно любое) файл
(ты можешь сам добавлять адреса)

Figure 1: An excerpt of the attack instructions found on a web site during the event.

- In general, the attacks can be described as Denial of Service (DoS) or Distributed Denial of Service (DDoS) attacks. Many well known methods were used, including ping flood, udp flood, malformed web queries, e-mail spam, etc.
- A few more complex attempts were made to hack into systems, for example using SQL injection. Some of these attacks met with success at non-critical sites.
- The targeted systems included web servers, e-mail servers, DNS servers and routers. Most visible to the public were the attacks against web servers.
- The targeted entities included the government, the president, the parliament, police, banks, Internet service providers (ISPs), online media, as well as many small businesses and local government sites.

- May 9th is an important date in this event, because that is when Russians celebrate victory over Nazi Germany. On many sites (including the example in Figure 1) the organizers called for an attack on that politically important date. The big attack wave anticipated for May 9th started shortly after 11PM local time on May 8th, however, suggesting that these attackers were on Moscow time.

2.2. Facts gained from the Estonian state procurature in January 2008

- As of January 2008, only one person has been convicted of carrying out cyber attacks in the spring of 2007. Dmitri Galuškevitš, a 20-year old student in Estonia was fined for organizing a DDoS attack against the website of a political party in Estonia. His conviction was possible because he committed the attacks from Estonia and therefore enough evidence could be collected.
- The Estonian State Procurature made "a formal investigation assistance request" to the Russian Supreme Procurature in May of 2007, in order to track down attackers residing in Russia. As of January 2008, this has not yielded any positive response, regardless of the fact that this type of cooperation is specifically "enumerated in the Mutual Legal Assistance Treaty" between Estonia and Russia.

3. Analysis

The analysis will attempt to find a plausible explanation for the cyber attacks that took place against Estonian information systems between 27th of April and 18th of May 2007. Due to the nature of the facts gathered in the previous chapter, the author will use qualitative analysis. Several hypotheses are considered:

- The event was a Russian information operation against Estonia

- The event was a false flag operation to frame Russia as the attacker
- The event was a spontaneous grass root level response to the policy of the Estonian government

This is not a complete listing, but the author feels that these three can be considered as the most probable explanations to the event.

3.1. Information operation

In this analysis, the author considers an information operation as the use of information and information technology to affect the decisions and actions of an opponent. In an article about possible Chinese strategies for invading Taiwan, Wu (2004) points out the possibility of using the information age equivalent of the concept of *people's war*. In the context of cyber attacks, this means that ordinary citizens of a state can be motivated to use the resources under their control to independently attack enemy systems in order to confuse the defenders. Amidst all the noisy and ill-coordinated attacks, more professional intrusions can then be carried out, supplemented with physical attacks to take out the command and control systems of the opponent. (Wu 2004) The beauty of people's war is that it provides near perfect deniability for the government or any other entity that is behind the attacks.

In order to consider this hypothesis plausible in the context of people's war, we need to show that:

- many people of varying skill levels took part in the attacks;
- the people who committed the attacks were externally motivated; and,
- the attackers received some form of support from the state.

Judging from the variety and volume of different attacks, it is likely that they were committed by many different individuals. The only person convicted of taking part in the attacks was shown to be responsible for an insignificant fraction of the attacks while further investigations have stagnated due to the lack of cooperation by Russian authorities. The activity in forums at the time of the attacks also indicates widespread interest in attacking Estonia. The attacks ranged from manually launching pings to botnet DDoS's to exploiting specific vulnerabilities in router software. Many of the detected attacks were described in detail on various Russian language forums and websites, which were easily available to those interested in finding a way to participate in the attacks. Most of these instructions were extremely simple to execute, thus making the prior experience of the attackers irrelevant. Therefore, we can say that many people of varying skill levels likely took part in the attacks.

It is also clear that the attacks were politically motivated because many of them contained a message related to the overall conflict surrounding the statue. The hostile rhetoric from various high ranking politicians in Russia were broadcast in the media and disseminated further in forums and web portals. On some of these forums there were open discussions about attacking Estonian systems or collecting resources for renting botnets. Taking the preceding factors into consideration, one can easily see that the attackers received encouragement from high ranking members of the Russian political elite.

The Russian government has consistently denied any direct involvement in the cyber attacks that hit Estonia in the spring of 2007. To the author's knowledge this claim is true. It is remarkable, however, that neither is there any proof of measures taken by the Russian government to mitigate the situation. The lack of cooperation in the Estonian investigation indicates that the Russian government is not interested in identifying the attackers and is therefore, in essence, protecting them. In other words, hostile rhetoric from the political elite motivated people to attack Estonia while nothing was done to stop the attacks. This silent consent, however, can be interpreted as implicit state support because without fear of retribution the attackers were free to target Estonian systems.

Assuming that this event was a result of a deliberate information operation, it is most likely tied with the larger political conflict that surrounded it. Since no entity has claimed responsibility for organizing the attacks, the author can only speculate as to the aim of this operation. In this case, the aim could be to unite the Russian people against a common enemy before the elections. Another possibility is to destabilize the Estonian society and to undermine the Estonian economy in an effort to weaken its ties to the European Union and the North Atlantic Treaty Organization. Yet another is a proof of concept on the digital people's war idea while supporting the overall political campaign surrounding the statue. At least in theory, several reasons can be found for conducting this type of operation.

If the cyber attacks were the result of an information operation, then one could argue that it was fairly effective. Large scale attacks were mounted against an independent state while no controlling entity (government or otherwise) has been identified. This would be an invaluable lesson for preparing for future conflicts. Therefore, this hypothesis can be considered plausible.

3.2. False flag operation

It has been suggested that the cyber attacks could have been a false flag operation. In other words, that the theoretical mastermind behind the attacks wanted to make it look like it was originating from Russia. While an interesting theory, it fails to explain the hostile statements of the Russian officials and the complete lack of cooperation on the investigations of the cyber attacks originating from Russia. In case of a false flag operation, it would be in Russia's interest to show the world that they were in fact not behind the attacks and better yet, to expose the entity that planned it. As a result, this hypothesis is implausible.

3.3. Grass roots response

Another theory is that the cyber attacks were nothing more than a wide scale, international, grass roots protest against the policies of the Estonian government. This would explain why no organization, agency or government has taken responsibility for the attacks. Unfortunately, this theory would require only spontaneous actions of the people while silent state support has already been demonstrated in previous sections. Once we admit the state as one of the partners in the protest, it is no longer grass roots or independent. Therefore, this hypothesis is implausible.

4. Lessons learned

One of the biggest lessons emerging from this event is that in a modern conflict, cyber attacks are becoming increasingly more common and dangerous. Any country with sufficiently well developed network infrastructure is vulnerable to these attacks. Primitive cyber attacks take very little time and effort to organize, while defending against them is becoming more and more difficult. Under the cover of the primitive and noisy attacks, more professional intrusions can be performed to gain a foothold for further attacks.

There are several problems with using the Internet as a field of battle by lone hackers, terrorist groups and states. First, the Internet spans the globe, thus a large scale attack is likely to influence innocent bystanders in other countries as well as the target country. Therefore, some of these attacks could be classified as terrorist activity, since they target civilian systems in the hopes of getting more attention from the press.

Second, the relative anonymity of the Internet allows for a near perfect deniability, as was the case in Estonia. All one has to do is either originate the attack from or route the traffic through a country that is not willing to cooperate. This makes it almost impossible to bring the attackers to justice, especially when considering the lack of common international legal grounds for these new types of attacks and conflicts.

Third, a new phenomenon is currently emerging that could change the concept of information assurance in a radical fashion. This phenomenon is the militarization of cyber space. Most systems today are built with lone hackers and script kiddies in mind. But militaries are moving into cyber space. What if all the nationally critical systems fall under a simultaneous concentrated cyber attack from thousands of professional, well trained and equipped cyber attackers? In a war scenario, these attacks would most likely be complemented with physical destruction at some key sites, as well as special operations troops capturing others. The author believes that this could be devastating to any country with a developed network infrastructure. Organized military resistance could be knocked out overnight, in theory.

5. Summary

The analysis of the cyber attacks that hit Estonian systems in the spring of 2007 is a difficult task due to the fact that a large part of malicious network traffic data is unobtainable. This, in turn, does not allow the investigators to pursue many of the people who committed the attacks. Therefore, the author used qualitative analysis of the known facts to provide an overall explanation for the event.

Of the three hypotheses considered, only one was determined plausible. The author concluded that the event can be explained as a Russian information operation against Estonia. Specifically, this event seems to match the digital version of the Chinese concept of people's war, where the government motivates people to attack its enemies by any means at their disposal. The digital version provides plausible deniability for the government, while in the case of this event the government can easily protect the attackers by refusing to cooperate with foreign investigators. This scenario illustrates the many dangers that come with using the Internet as a battle space.

It should be noted that this analysis does not *prove* that there was an information operation due to lack of evidence from the Russian authorities. Instead, the conclusion is considered *plausible* and in line with the available facts. If the Russian authorities were to release the necessary technical evidence, a more thorough quantitative analysis could be conducted, which could lead to the attackers.

References

International Centre for Defence Studies (ICDS) (2007) "Moskva käsi Tallinna rahutustes. Rahvusvahelise kaitseuuringute keskuse kiirülevaade 7. mail", *Sõdur*, No 2, pp 4-8. *(Moscow's Hand in the Tallinn Riots. A Quick Overview by the International Centre for Defence Studies on 7th of May)*
Ottis, R. (2007) *Analysis of the Attacker Profiles in the 2007 Cyber Attacks Against Estonia*. Unpublished MSc dissertation, Tallinn Technical University, Tallinn.
Wu, C. (2004) "An Overview of the Research and Development of Information Warfare in China." In Edward Halpin et al (eds.) (2006) *Cyberwar, Netwar and the Revolution in Military Affairs*. Palgrave MacMillan, Hampshire, pp 173-195.

Electronic Activism: Threats, Implications and Responses

Allen Wareham[1] and Steven Furnell[1, 2]
[1]University of Plymouth UK
[2]Edith Cowan University Perth Australia
Originally Published in the Conference Proceedings of ECIW 2008

Editorial Commentary
In addition to the physical threat of terrorist activities, the utilisation of computer systems and services to promote a further threat has also become a point of significant interest. While often dubbed 'cyber terrorism' by Government groups and media agencies, a more accurate description would often be 'electronic activism' and the associated risk has seemingly increased due to widespread adoption of the Internet.

This paper discusses the various aspects of activism online, utilising analysis of cited attacks to examine the methods and impact of direct pressure at both the corporate and national level. The discussion reviews two separate acts, against national and corporate entities, in order to highlight the scope and capability of electronic activism. The former is based upon events in Estonia, in which a series of Denial of Service attacks were targeted against Government services and infrastructure. The corporate example is based around Huntingdon Life Sciences, which was the target of a joint campaign involving both physical and technical threats to its operations. Using the UK as an example, the paper then proceeds to consider the legal frameworks in place to deal with e-activism and wider cyber terrorist threats.

Allen Wareham and Steven Furnell

Abstract: In addition to the physical threat of terrorist activities, the utilisation of computer systems and services to promote a further threat has also become a point of significant interest. While often dubbed 'cyber terrorism' by Government groups and media agencies, a more accurate description would often be 'electronic activism' and the associated risk has seemingly increased due to widespread adoption of the Internet. It is relevant to consider what has actually been seen in relation to e-activism, as a concept and with reference to related examples, in order to assess the nature of the threat that it poses. This paper discusses the various aspects of activism online, utilising analysis of cited attacks to examine the methods and impact of direct pressure at both the corporate and national level. The discussion reviews two separate acts, against national and corporate entities, in order to highlight the scope and capability of electronic activism. The former is based upon events in Estonia, in which a series of Denial of Service attacks were targeted against Government services and infrastructure. The corporate example is based around Huntingdon Life Sciences, which was the target of a joint campaign involving both physical and technical threats to its operations. Using the UK as an example, the paper then proceeds to consider the legal frameworks in place to deal with e-activism and wider cyber terrorist threats. The discussion shows that while the more aggressive forms of activism could be charged under anti-terrorism laws, such reactive measures may not deliver the required solution in all cases, especially when the attacker feels that they acting on a 'higher authority' based upon moral, ethical or religious grounds.

Keywords: Activism, cyber terrorism, hacktivism, intimidation, influence

1. Introduction

The issue of cyber terrorism has grown considerably in the public eye since the events of 9/11, with the issue of Internet-based threats receiving further attention as a consequence. However, in addition to the potential for hosting a direct attack, the Internet has also become a recognised platform for related problems. For example, a variety of websites have been established to promote extremist views, and consequently provide a potential recruiting mechanism for radical groups (BBC, 2008). The motivation of such threats is not purely limited to religious idealism; the considerations of moral, ethical and political standpoints have their own parts to play, with examples of such efforts giving rise to this current paper.

The presence of activism on the Internet is a problem that requires more than simply analysing the potential technical attack methods, with considerations on social factors, legal issues and present legislation, the current UK defence policy to electronic attack and numerous other factors. This discussion paper examines a series of core areas and examples in order to draw a more informed opinion. The paper will then conclude through the suggestion of potential areas of further study, in order to try and pinpoint an effective and realistic solution to the malicious use of information systems.

2. The reach of electronic activism

Since its inception, the Internet has been a natural conduit for the exchange of ideas, evolving from the early ARPANET into a global phenomenon. The widespread acceptance of the web has created a universal resource for business, community support, education, finance and Government; a link to the outside world which is becoming increasingly important in everyday life. Moreover, the public nature of the Internet provides a number of opportunities for any group who wishes to centralise their activities online, both in terms of services and environment:

- *Operations Support* - The ability to build and maintain the necessary 'command network' from which to operate as a cohesive group, a consideration especially important when considering the large global geographical and transnational spread of potential large-scale efforts.
- *Anonymity* - The ability to "hide in the anonymity of cyberspace" (Jones, 2005), allowing potential planning, coordination and actions to be overlooked until the goal has been completed.
- *Personal 'Safety'* - The ability to use considerations of anonymity from which to operate, a potential benefit and incentive to less 'driven' orchestrating groups or lone individuals.

- *Publicity* - The ability to publicise objectives, viewpoints and motivations for the purpose of informing, persuading and the incitement of propaganda, coupled with the maintaining of a desired 'public image' to validate or support operations and ideals.
- *Financing* - The ability to maintain continued operations through continued financing, through both donations and the utilisation of legal or illegal credit transactions

Notably, these opportunities are offered to both legitimate groups, as well as those who may otherwise find it difficult to operate due to the risk of exposing potentially illegitimate activities. This itself becomes a problem when considering the 'no questions asked' state of many Internet services, largely stemming from an environment where the free flow of ideas is actively encouraged without any particular national restriction.

2.1. Establishing and maintaining communities

The Internet has a substantial capacity to further and support community elements in commerce, entertainment and academia, with social networking sites and privately managed communities forming to focus on innumerable topics of interest. The majority of these groups offer a benign and harmless outlet for the free exchange of ideas. However, the online community can offer its own potential dangers, and the ways in which services can be utilised depend largely on the needs of the group in question. For many the ability to publicise information regarding their goals and aspirations, if only in a few brief paragraphs, is a basic component of web presence; perhaps even their main or only point of contact with the 'outside world'. A further necessity for many established communities is a need to authenticate users in order to view more pertinent information of interest, commonly in the form of membership or some other standardised process so that access to key services can be achieved. The same considerations also are in existence when considering the issue of activism, the balance of organisation and coordination processes with the presentation of intended material for the general public, whilst ensuring that sensitive and potentially incriminating information is secured. The overall cost provisions in terms of infrastructure are for the most part negligible, with tools such as IRC, phpBB and instant messaging systems providing flexible and free options for communication. The introduction of secure data using such tools as PGP adds further credence to the management of 'closed' community environments, systems which can cause significant headaches for policing and the security services. As a case in point, in 2006 the UK's Serious Organised Crimes Agency (SOCA) found itself essentially foiled by the use of encryption during the raid on an ID theft ring, the estimated time to break the encryption "taking 400 computers twelve years to complete" (Espiner, 2006).

2.2. Influence and intimidation – the tools of the trade

The ability to influence the viewing audience effectively is a significant issue when considering the nature of electronic activism. The ability to 'prove' or 'disprove' specific information, as well the legitimising of actions, provides activists a certain amount of validity to justify intent, such as the committing of actions that may otherwise be viewed to be entirely inappropriate in the public spectrum. Hutchinson (2007) explains the basic considerations of *propaganda* and *persuasion* in the common presentation of ideas, the subtle difference between the two methods highlighted in Table 1. The processes of each are self evident in numerous online publications and on a range of media, with services such as *YouTube* providing an ideal host to portray supportive material to capture the global audience. The ability to create 'realities' through the submission of convincing and relevant material allows for the swaying of public opinion, enhanced further by the provision of community groups to strengthen given perceptions. Arguably this is not a new concept; indeed the reinforcement of values, ethical principles, ideology and given arguments can be seen in virtually every current mainstream religion and Government across the globe. The main difference we can see when considering electronic activism is that the adoption of full media web services allows for the presentation of emotionally driven material that can both support activist efforts and separate the target from perceived legitimate protection.

Table 1: Classifications of influence

Influence Type	Approach	Influence Strategy
Propaganda	Wide-scale	Sociological principles reinforcing cultural or social values
Persuasion	"Personal"	Psychological principles and arguments

2.3. Electronic activism in offensive scenarios

Whilst the base mechanics and provisions for the co-ordination and application of electronic attacks may be covered, the review of specific cases can often be quite difficult. This is largely due to the unwillingness of targets to be open regarding potential lapses of security, more so the consideration that they may be a continuing viable target or that the target's interests may have been harmed (such as the divulging of customer information or technical product data). Due to the difference in scale between a corporate and a national attack the following examples highlight a case for each respective instance.

An example that highlights the potential threat of a national attack is provided by events in April and May 2007, which demonstrated the potential impact of activism on a national scale. The relocation of a Soviet war memorial in the City of Tallinn sparked a high profile conflict with both the Russian Federation and ethnic Russians living within Estonia; the apparent catalyst for the later attacks on Government services and infrastructure. The attack sustained for a number of weeks, with Table 2 emphasising the range frequency of attacks committed during a monitored period.

The thought that Estonia is heavily dependent on the internet to support government, civil and financial institutions is indeed concerning, considering that Estonians "pay taxes online, vote online, bank online (and that) their national ID cards contain electronic chips" (Applebaum, 2007). This means that an effective *Denial of Service* attack increases in the potential effect on a target as the sophistication of the target increases (a concerning trend due to a similar embracing of technology in the UK).

The sheer fact that the Estonian government brought the issue before NATO as a legitimate attack on its sovereignty highlights the validity and seriousness of the incident, not merely as an annoyance but most certainly as a full blown attack in its own right. The response from NATO Secretary General was that of voiced condemnation over the incident (Estonian Government, 2007), although further action was not clearly identified.

Table 2: Attacks against Estonian websites (Nazario, 2007)

Attacks	Destination	Address or owner	Website type
35	"195.80.105.107/32"	pol.ee	Estonian Police Website
7	"195.80.106.72/32"	www.riigikogu.ee	Parliament of Estonis website
36	"195.80.109.158/32"	www.riik.ee	Government Information website
		www.peaminister.ee	Estonian Prime Minister website
		www.valitsus.ee	Government Communication Office website
2	"195.80.124.53/32"	m53.envir.ee	unknown/unavailable government website
2	"213.184.49.171/32"	www.sm.ee	ministry of social affairs website
6	"213.184.49.194/32"	www.agri.ee	ministry of agriculture website
4	"213.184.50.6/32"		unknown/unavailable government website
35	"213.184.50.69/32"	www.fin.ee	ministry of finance website
1	"62.65.192.24/32"		unknown/unavailable government website

Animal research and testing has long attracted the attention of animal rights activists on both a national and transnational level. One such group, the *Stop Huntingdon Animal Cruelty* activist group (Affiliated with the international group PETA), has been involved on numerous occasions with illegal activity, primarily in terms of physical actions such as the storming of target offices and direct intimidation methods of targeted personnel.

Analysis of legal case reports from March and May 2004 highlight a variety of offences against the HLS facility and its personnel, with the documents demonstrated a joint campaign of both physical and technical threats to continued operations. The usage of phone, fax and email blockades, the interruption of mobile phone services and the harassment of affiliates were identified as the primary forms of technical attack, and although extremely basic, these methods proved enough in conjunction with physical efforts to drive away a number of investors. Indeed, the intended "impacting (on) Huntingdon's bottom line" (QB, 2004) as voiced by one of the defendants was an important part of many of the highlighted public comments, with the viewpoint of potential prison sentencing being "a small price to pay". (QB, 2004) This highlights understandable concerns over legal steps when attempting to deal with persistent or repeat offenders, especially when considering that one of the defendants in the reviewed cases highlighted legal action as being "laughable because we will find a way around it" (QB, 2004).

3. National security and defence measures

Looking at the UK context as an example, the primary recourse against actions committed has been that of the law, specifically in the case of terrorism (and thereby potentially most applicable to defined activism) three laws in particular, namely the *Terrorism Act 2006*, the *Prevention of Terrorism Act 2006*, and the *Anti-Terrorism, Crime & Security Act 2001*. Table 3 highlights the powers that these deliver, with the Terrorism Act in particular referencing *Internet* activities. This is a significant tool when combating potential aggressive forms of electronic activism, referencing key issues that (especially in terms of the HLS example) can be applied to situations where a particularly aggressive threat may be present. Further to the list in the Table, the RIPA Act 2000 further requires the handover of all keys and information relating to encrypted data for investigation, with a potential prison sentence for up to 5 years should there be any issues of National Security (Home Office, 2007). As highlighted by the example of encryption employed against SOCA investigations, this may not be a large enough incentive should the potentially discovered information lead to a greater sentence. However, this covering period provides a mechanism to legally charge and detain potentially disruptive or dangerous elements.

Table 3: An overview of key UK anti-terrorism laws

Act	Focus
Terrorism Act 2006	Designed to combat: Planning of Terrorist Acts Encouragement of Terrorism Dissemination of Terrorist Publications The Training of Terrorists
Prevention of Terrorism Act 2006	Designed to: Impose sanctions on specific suspects including the introduction of control orders Note: carried out in accordance to the EHCR and authorised directly by the Home Secretary
Anti-Terrorism, Crime & Security Act 2001	Designed: Combat Terrorist funding operations Extend police powers

The overall defence of UK infrastructure and interests is largely under the jurisdiction of the Home Office, assisted by independent groups such as the Joint Intelligence Committee (JIC) in the development of a suitable overall strategy. This structure is illustrated in Figure 1.

The four core groups that deal with the main body of the UK civil infrastructure and assets (CSIA, 2007) are as follows:

- CESG: an arm of GCHQ which primarily focuses on the protecting of national information systems,
- CSIA: a cabinet body catering primarily to the advisory of
  government groups and departments,
- CPNI: an arm of MI5 which focuses on protecting the UK's
  infrastructure from Electronic Attack
- SOCA: a progression from the now defunct National High
  Tech Crime Unit which deals with primarily high level crime
  / high impact crime.

These groups work together in order to protect against potential attack, as well as to ensure that information policy is maintained. In matters regarding the protection of military interests the Ministry of Defence operates its own defensive resources. Below the upper strata of national protective elements operate the various CERT and CSIRT teams that collaborate to protect key areas such as banks, businesses, commercial infrastructure, educational facilities or less significant government concerns. These operate either independently or in mutually beneficial co-operative organisations such as FIRST. Overall this gives a reasonable level of infrastructural security, operating on a tiered basis in which each party can consult others for mutual support, development, education and protection, whilst alleviating stress on Government groups. Collectively, this offers a good level of defence against focused Denial of Service attacks (one of the standard methods of attack) should communication between the various groups and Government bodies be sufficiently managed. However, short-term or intermittent attacks where a reactionary effort may take time or minor transnational 'nuisance' efforts may be still a difficult problem to face.

Home Secretary

Indep. Groups	Home Office	EU
advisory	policy and decision maker	advisory

CESG	CSIA	CPNI	SOCA
arm of GCHQ	arm of the cabinet	arm of MI5	arm of the police

Governing Group CERT Authority

BPDSAC · BTCERTCC · BTSBS · BUNKER · CSIRTUK · DAN-CERT · DANTE · E-CERT · JANET-CERT · MOD-CERT · OxCERT · Pentest · Q-CSIRT · RBS-G-ISIRT · RM CSIRT · SKY-CERT · UNIRAS

Indep. CERT / Non CERT

example: FIRST CERT group

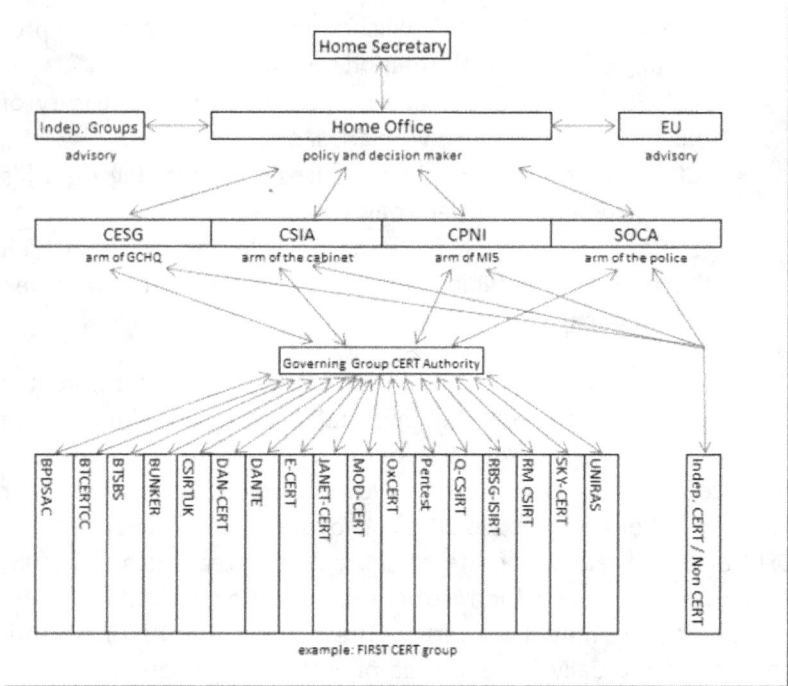

Figure 1: A basic overview concerning UK defence

4. Discussing electronic activism

Overall, the considerations of activism highlight an embrace of basic attack methods in order to *deny* the target its communication capability, either in conjunction with defamation attacks as used in the Estonia, or by the usage of intimidation as prescribed by the events of the HLS case. The utilisation of electronic activism as a communication method provides opportunities for the conversion of others to a similar way of thinking, facilitating both potential recruitment as well as an escalation of the impact of activities due to greater public support. The natural qualities of the internet and the provision of free products and services mean that implementations for activist groups require little financial input, with the potential of utilising financial services such as *Paypal, Western Union* and even online virtual environments such as MMOGs (Massively Multiplayer Online Games) and *Second Life*. The nature and protection afforded by authenticated 'closed' community environments means penetration by law enforcement can be difficult, and even further impacted by the possibility of free encryption methods.

The more aggressive forms of activism can be charged under anti-terrorism laws as discussed above when highlighting National Security and defence measures, with national infrastructure measures in place to counteract any wide-scale attack. However, these methods are for the most part reactionary responses rather than proactive methods, essentially waiting for the activist to make the first move. A further consideration when employing the law is that legal jurisdiction lies very much with the country from where the action was carried out, leaving significant concerns for transnational attacks. We have identified that legal consequences may not be a silver bullet to activist attacks, especially when the attacker feels that they report to a 'higher authority' based upon moral, ethical or religious grounds. The issue is compounded when considering that the previous head of MI5 Dame Eliza Manningham-Buller discussed the nature of the UK's proactive surveillance activities, citing a lack of manpower in relation to the threat (BBC, 2006). Following the bombings of 2005, the question over the sacrificing of certain civil liberties highlighted the difficulty in balancing measured security with effective security over the general 'terrorist' threat. The activist could in fact claim an abuse of human rights in terms of personal freedoms and privacy, meaning that otherwise effective methods could essentially be out of bounds. This raises the age-old discussion of the needs of the individual in relation to the needs of the state.

An issue that was identified when comparing the corporate and national examples was that although technical solutions can to a degree be provided, the issue of human-orientated attacks is a far harder issue to combat. The problems of intimidation and threats, although far smaller in terms of the potential target radius, can negate common methods of protection. With the focus of the attack on the user, the threat of disseminating personal information to friends, family and business relations in order to invoke an emotional response is common-place; one that was used against key directors and target business partners within the HLS example. With the impracticality of segregation and the screening of emails and similar forms of contact (one of the only real defences against the human side of activism), an effective solution would seem to require a blend of protective considerations:

- The implementation of an effective security policy for staff, assets and information systems at both a corporate and government level,
- The confirmed civil and government enforcement of contraventions regarding either security breaches or acts of an activist nature (including the prosecution of trans-national transgressors),
- The assigning of responsibility to service providers for the usage of the services they provide.

The nature of defence commonly requires a blanket approach when considering the nullification of potential threats, with an 'all or nothing' approach highlighting the inherent risk of defending in some areas whilst leaving others open for exploitation. The requirement to defend against both technical and human-based attacks is a mandatory necessity, since the shift in terms of requirements to match lacking defensive measures is relatively minor. The effort to 'cover all the bases' may prove costly, however as in all business scenarios the risk of not putting in sufficient security measures can cost far more.

One of the best potential solutions to the issue of electronic activism is the alerting of consciousness and the enabling of disinhibitors; key principles initially defined in a Home Office report on future net crimes. The ability to influence the general public and demonstrate that the *action* of vigilantism (even if justified by some moral, ethical or religious concept) is wrong will potentially offer a solution, either by directly influencing the potential activist or by inciting peer pressure through others to search for a more peaceful solution to grievances. As such, it is somewhat ironic that the same routes available for the conduct of electronic activism may also offer a potential channel for reducing the problem by promoting alternative views.

5. Conclusion

Although there are strong indications that national infrastructure is well-maintained and provided for, the issue of national security continually being on the 'back foot' provides numerous opportunities for exploitation. The threat of human-orientated attacks allows for effective erosion of a target's support, as well as offering potential sources of unrest by inciting propaganda/persuasion fuelled discourse through 'legitimate' information sources (through the creation of *realities*). This presents a significant threat at the personal and corporate level, with implications if carried out on a large enough scale to even effect national-level engagements. The law may help combat the issue of misusing information systems as both a deterrent as well as through incarceration, but it is by no means the final solution against more dedicated activists, especially who live outside the national boundary within a legally or politically unsympathetic county. The threat of effective attack methods is compounded by inexpensive necessary resources being freely available to a global population, more attuned to the usage of technical web systems and applications.

The potential of activism is required by all involved parties from a personal, corporate and national perspective, with the proper level of communication to help facilitate the recognition and combating of an authenticated attack. The re-evaluation of the relevance of security concepts at each level could help support this, especially when reinforcing the required responsibilities of those at every level.

The enabling of the general public to regard the malicious effects of electronic activism as a very real and human consequence could bring about the necessary change in public opinion to help combat and isolate the risk of national threats.

A number of suggestions for further areas of study lie within both academic and government domains. The study of effective communication types within activism, and an examination of the effectiveness of available malware resources and standard tactics could highlight the base effectiveness of electronic activism, allowing for methods to be potentially developed to pre-emptively tackle activist groups before they become a problem. The controlled penetration testing of government systems using freely available resources could highlight weaknesses in infrastructure systems and policy. The study of the potential effect of adverse cultural change in regards to sustained activist efforts, or the evaluation of propaganda and persuasion techniques in bringing about positive change to negate the support of activism could help forecast and even nullify UK-based threats in the future, although without a combined global effort trans-nation threats will always exist.

References

Applebaum, A (2007) "For Estonia and NATO, A New Kind of War", The Washington Post, 22 May 2007, http://www.washingtonpost.com/wp-dyn/content/article/2007/05/21/AR2007052101436.html (accessed 23/01/2008).

BBC (2006) "MI5 tracking '30 UK terror plots'", BBC News online, 30 November 2006, http://news.bbc.co.uk/2/hi/uk_news/6134516.stm (accessed 23/01/2008).

BBC (2008) "Smith targets internet extremism", BBC News online, 17 January 2008, http://news.bbc.co.uk/1/hi/uk_politics/7193049.stm (accessed 23/01/08).

CSIA (2007) "Key organisations", Central Sponsor for Information Assurance, Cabinet Office, http://www.cabinetoffice.gov.uk/csia/key_organisations (accessed 23/01/2008).

Espiner, T. (2006) "ID theft gang thwarts police with encryption", ZDNet News, 18 December 2006, http://news.zdnet.co.uk/security/0,1000000189,39285188,00.htm (accessed 23/1/2008).

Estonian Government (2007) "NATO Secretary General to the President of the Republic: the alliance supports Estonia", Government Communication Office Briefing Room. 3 May 2007.
http://www.valitsus.ee/brf/?id=283225 (accessed 23/01/2008).

Home Office (2007) "Frequently Asked Questions", Regulation of Investigatory Powers Act, http://security.homeoffice.gov.uk/ripa/encryption/faqs/ (accessed 23/01/2008).

Hutchinson, W. (2007) "Using Digital Systems for Deception and Influence", in Proceedings of the International Symposium on Human Aspects of Information Security & Assurance (HAISA 2007),

Plymouth, UK, 10 July 2007, pp79-86.

Jones, A. (2005), "Cyber Terrorism: Fact or Fiction", Computer Fraud and Security, June 2005, pp4 – 7.

Morris, S. "The future of net crimes now: part 1 – threats and challenges", Home Office, www.homeoffice.gov.uk/rds/pdfs04/rdsolr6204.pdf

Nazario, J (2007) "Estonian DDoS Attacks: A summary to date", 17 May 2007, Arbor Networks,
http://asert.arbornetworks.com/2007/05/estonian-ddos-attacks-a-summary-to-date (accessed 23/01/2008).

QB (2004) Chiron Corpn Ltd and Others v Avery and Others, EWHC 493

www.ingramcontent.com/pod-product-compliance
Lightning Source LLC
Chambersburg PA
CBHW050528270326
41926CB00015B/3125